# Goat Farming

## A Comprehensive Guide to Breeding, Health, Feeding, Products and Many Other Important Aspects of Goat Farming

### By

### Various Authors

# Contents

## Choosing a Goat

Starting Right with Milk Goats. Helen Walsh........……...………*page* 1

Family Goat-Keeping. W O'Connel Holmes...…….…….....……*page* 12

Goat Keeping - Written for the National Federation of Young Farmers' Clubs with the Assistance of the British Goat Society. Anon.............................................................………....…...…*page* 17

## Breeds of Milk Goats

Improved Milk Goats - A Guide for Breeders, Dairymen and Exhibitors. Will L Tewalt...….……...…………………………*page* 31

## Breeding Goats

Family Goat-Keeping. W O'Connel Holmes...………...…….…..*page* 85

Improved Milk Goats - A Guide for Breeders, Dairymen and Exhibitors. Will L Tewalt..........................................…....……*page* 88

Practical Angora Goat Raising. Anon.............................…..*page* 92

Starting Right with Milk Goats. Helen Walsh.………....…...…*page* 107

# Feeding Goats

Improved Milk Goats - A Guide for Breeders, Dairymen and Exhibitors. Will L Tewalt.................................................*page* 120

Starting Right with Milk Goats. Helen Walsh....................*page* 128

Goat Keeping - Written for the National Federation of Young Farmers' Clubs with the Assistance of the British Goat Society. Anon.................................................................*page* 147

Family Goat-Keeping. W O'Connel Holmes....................*page* 152

Practical Angora Goat Raising. Anon..........................*page* 164

# Kidding

Family Goat-Keeping. W O'Connel Holmes.....................*page* 167

Goat Keeping - Written for the National Federation of Young Farmers' Clubs with the Assistance of the British Goat Society Anon..................................................................*page* 179

Starting Right with Milk Goats. Helen Walsh..................*page* 189

# Goat Health

Family Goat-Keeping. W O'Connel Holmes.....…...……*page* 196

Practical Angora Goat Raising. Anon............…….….….…*page* 201

Goat Keeping - Written for the National Federation of Young
Farmers' Clubs with the Assistance of the British Goat Society.
Anon...............................................................................*page* 205

Improved Milk Goats - A Guide for Breeders, Dairymen and
Exhibitors. Will L Tewalt..............................…...………...…….*page* 211

Starting Right with Milk Goats. Helen Walsh .....….…....…..*page* 217

# Goat Products

Family Goat-Keeping. W O'Connel Holmes.....…...…….……*page* 223

Goat Keeping - Written for the National Federation of Young
Farmers' Clubs with the Assistance of the British Goat Society
Anon.........................................................….….……..…..*page* 229

Improved Milk Goats - A Guide for Breeders, Dairymen and
Exhibitors. Will L Tewalt....................................…………....*page* 233

Starting Right with Milk Goats. Helen Walsh...............*page* 250

Family Goat-Keeping. W O'Connel Holmes......................*page* 293

Improved Milk Goats - A Guide for Breeders, Dairymen and Exhibitors. Will L Tewalt.........................................*page* 297

Practical Angora Goat Raising. Anon..........................*page* 299

Family Goat-Keeping. W O'Connel Holmes..................*page* 308

## The Goat Farming Business

Improved Milk Goats - A Guide for Breeders, Dairymen and Exhibitors. Will L Tewalt..........................................*page* 311

Starting Right with Milk Goats. Helen Walsh..............*page* 325

# What Breed to Buy?

Don't rush off now and buy the first goat you see, much as you would buy a bag of salt. Take it easy, and first learn something about goats.

By the way, learn that the proper term for a female goat is doe, for a male, buck, and if you would have the goodwill of goat lovers don't refer to them as nanny goats and billy goats. Goat people hate this just as you would hate hearing your children called brats.

Noah was a good man, and he was also a wise man. When he equipped the Ark he put into it two animals of each species. His primary purpose, of course, was propagation, but he also knew that no animal can be happy alone. You will find that your goat will be happier with a companion and will make fewer demands on your attention. You might buy two milkers, if you need that amount of milk, or you might get a milker and a yearling—a debutante, so to speak—one just reaching maturity. Goats are seasonal milkers, milking usually for eight or ten months. By breeding the yearling so that she will have a baby kid and begin to give milk—which is called freshening —before the milker goes dry you can keep a supply of milk throughout the year. Or you might buy a milker and a kid. But be sure it is a doe kid.

After you have held a family council and decided that you want goats, visit the goat breeders in your section of the country, look over their stock and talk goats with them. Most goat raisers love their animals and love to talk about them and, generally speaking, will give you freely of their knowledge. It is preferable to buy your goats from such a source, for these people know how to feed and care for the goats, their animals have been tested for TB and Bang's disease, just as cows are tested, and you will know that the milk is safe for use.

You will find, as you go about, that each breeder specializes in raising one or more types of goats, and that there are four popular breeds. These breeds are Toggenburg, Saanen, Nubian, and Alpine.

Toggenburgs, originating in Switzerland, are almost uniformly some shade of brown, with a white stripe at either side of the face from eye to muzzle, a white area around the tail and white hocks. There are, however, occasional variations toward grey coloring, or even spots of white on the coat infrequently. Their faces are slightly concave between the eyes—"dished" they are called—their ears are prick or stand-up, their coats either short or long haired (usually long haired in the buck), and does as well as bucks have beards.

Saanens, also from Switzerland, are large, white or cream white animals with dished faces and prick ears, and are either short or long haired. Occasionally a Saanen will show spots of grey or touches of black, but the goal of the Saanen breeders is the pure white animal. The Saanen doe also is bearded.

Anglo-Nubians, commonly referred to as Nubians, may easily be recognized by their large, drooping ears and humped (Roman) noses. They were originated by the British by crossing the native goat with bucks from Africa. They are always short haired, but of no specific color. They may be of various shades of brown, black, white, cream, buff, grey or combinations of these colors.

French Alpines are large animals with cone-shaped ears and slightly concave faces. They too may be white, black, brown, grey or combinations of these colors. In the cou blanc type the

Famous bucks and does who helped establish their breeds here and from whom many present day goats are descended.
*Upper.* Prince Bismarck from Switzerland and Polly Mac. (Toggenburgs). *Lower.* Lunesdale Spurius Lartius from England and Shirley Rona, his daughter, from Canada. (Nubians)

front part of the body is light in color, shading into black hind quarters. Chamoisees are brown with black on head, legs and back; sundgaus black with white on face and under-part of body. Their hair is sleek and short.

Rock Alpines, similar in appearance to the French Alpines, were originated in this country by Mrs. Mary Rock, who crossed the French Alpine with Toggenburg and Saanen stock. The Rock Alpine is the only breed of American origin.

A purebred is an animal of unmixed breed, descended from stock that traces entirely within the blood lines of the foundation stock of the breed. A grade is an animal one of whose parents is a purebred and the other a scrub or an animal containing a considerable proportion of the blood of the same breed as the purebred parent. If the dam is purebred the offspring derives her breed classification from either parent. A doe resulting from the breeding of a registered buck with a doe not purebred takes her breed classification from that of the sire. For example, a doe kid born to a doe of mixed breed bred to a registered Toggenburg buck would be ½ Toggenburg; the next generation would be ¾ Toggenburg; then ⅞ and so on. This method of breeding each generation to a registered sire of the same breed is known as "grading up" and results in a fairly rapid improvement of stock.

There are two associations for the registration of goats, established to promote and improve the dairy goat industry in this country. The American Goat Society registers only purebreds. The American Milk Goat Record Association registers purebreds and also records as grades does or doe kids that are up to 15/16 of a particular breed. When the 31/32 generation is reached the animal may be registered in this association as "American" of the breed, such as American Toggenburg, American Nubian, etc.

Membership in these associations costs a small fee and yearly dues, and both associations register animals of non-members as well as of members.

If you purchase a purebred goat be sure that you secure her registration papers at the time of sale. Sometimes the inexperienced buyer fails to do this and finds later (as I did) that the seller "can't find the papers," and the socalled purebred can't be registered, nor can her kids, except as grades. Here, again, your protection lies in buying from a reliable breeder, for he has a reputation to maintain and his statements can usually be depended on.

Don't, however, be swept off your feet at the mere thought of papers with your animal, for unless the papers cover a good animal from productive ancestors their chief value is historic.

A few more bucks and does.

*Upper.* Panama Louise from Switzerland and Panama Prince Fribourg. (Saanen) *Lower.* Le Poilu and Molly Crepin—two of the several bucks and does used to develop the Rock Alpine breed.

Any breeder with a registered animal to sell will give you a transcript of its genealogy and tell you something of its history. Also there is much valuable information on individual animals and blood lines to be found in the trade publications—*American Goat News, Dairy Goat Journal, Better Goatkeeping* and *The Goat World*—and in the writings of various goat breeders who have contributed articles to these magazines. Mr. J. S. Fetter of Coldwater, Ohio, for example, has written exhaustively on the subject of Saanens and Saanen importations. Mr. E. S. Thompson of Bristol, Pa., and Mrs. Evelyn Latourette, Estacado, Oregon, have covered Toggenburgs, and for information on these two Swiss breeds there is the Swiss Goat Club of which Mrs. Latourette is secretary. On Nubians Mrs. Carl Sandburg of Flat Rock, North Carolina, and Mr. Lyle Hulbert of Rome, New York, have contributed largely of their knowledge and on French Alpines there are the writings of Mrs. F. N. Craver of El Paso, Texas. On Rock Alpines there are the records of Mrs. Mary Rock, the founder of the breed.

Even if you are purchasing a grade goat you can learn much about its heredity by investigating the history of its sire.

The following data on breed production is interesting. However, be sure to bear in mind that there is great individual variation in animals.

The 1945 Advanced Registry reports, as summarized in March 1946 *Better Goatkeeping*, show:

The average tested French Alpine gave 1985.3 lbs. of 3.7% milk.

The average tested Nubian gave 1626.5 lbs of 5.02% milk.

The average tested Rock Alpine gave 1950.0 lbs. of 4% milk.

The average tested Saanan gave 2325.9 lbs. of 3.59% milk.

The average tested Toggenburg gave 1902.3 lbs. of 3.18% milk.

# How to Buy a Good Doe

WHEN you go to look at a goat, notice the general appearance of the animal. Is she alert, clear eyed, her coat shiny? If the answer is "yes" it should be an indication of good health. Steer clear of the goat who is "nice and fat," for while a dairy goat should not be thin and gaunt, she should not be of the chunky, beef-cattle type.

Notice the general formation of her body-conformation. See that her hind legs are spaced apart sufficiently to allow for development of a good-sized udder, which should be firmly attached, not sagging, with teats that can be easily grasped. Milking a goat can be a pleasure, but it is far from it if she has a generous udder with small teats. See that the animal has a well rounded "barrel" which shows capacity for digesting quantities of bulky food such as hay. If you can be on hand when she is milked notice if her udder shows an appreciable shrinkage in size after she has been milked. It should; otherwise you may suspect that the good-sized udder isn't well filled with milk, but consists mostly of muscular tissue.

If you are buying a yearling or a kid and her mother and sire are present, look them over so that you may know what to expect when the youngster matures. In my very early days I was on the point of buying a young doe because she had blue eyes—the only time I had seen a blue-eyed' goat. Then I saw the mother who, although purebred, was very small and had been bred prematurely. Common sense told me that much shouldn't be expected of the kid and although it was hard to resist the blue eyes I reluctantly let common sense rule the day.

If the person from whom you purchase operates a dairy, you will undoubtedly find that his goats have been tested for tuberculosis and Bang's disease, or brucillosis. If they have not been tested, it is well to have these tests made, for, although neither disease is often found in goats, your source of raw milk needs to be above question as to its wholesomeness.

What should you pay for a goat? Prices vary in different sections of the country and at different seasons of the year. Because the large percentage of goats freshen during the late winter and early spring, winter milkers are not easy to obtain, and a doe giving milk during the winter months, if she can be purchased, will bring a price somewhat higher than she might at other seasons when more milkers are available. You will find, however, that a fairly uniform price will prevail in your section which may vary from $25 to $75 for grades, $40 to several hundred for purebreds.

How much milk should you expect from your doe? According to the U. S. Department of Agriculture, a goat giving 2 quarts a day is a good milker, one giving 3 quarts excellent. For advanced registry tests a two-year-old doe (which means a doe on her first freshening) is required to show an average milk production of approximately 2½ quarts a day over a period of 10 months. This is a sort of roll of honor registry and means that the goat who earns her letters AR is better than average. Because you are accustomed to think of milk in quarts, as you buy it, I am giving the figures in approximate quarts, but actually the milk is weighed—a pint equalling approximately a pound. And the requirement is 1500 pounds of milk or 52 pounds of butterfat over a test period of 305 days, allowing 60 days' rest period before kidding without test, during which time the doe normally should be dry. If these

Daily Production and Fat Content of Goat's Milk

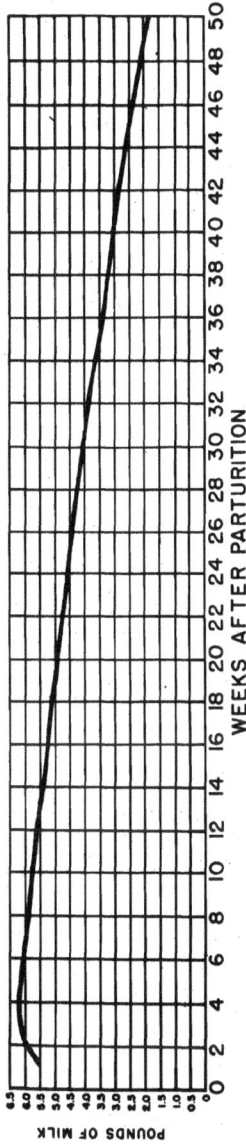

EFFECT OF PERIOD OF LACTATION ON THE DAILY PRODUCTION OF MILK AS RE-
PRESENTED BY COMPOSITE SAMPLES OF THE MILK OF SAANENS AND TOGGENBURGS

Maximum milk production ordinarily increases after each freshening until the fifth to eighth year when a doe is at her peak.

9

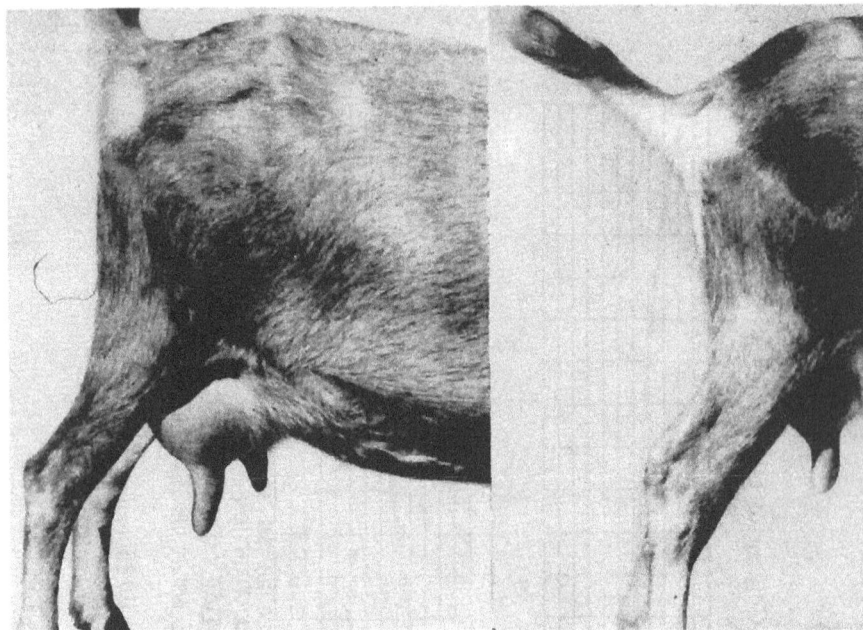

*Left.* Goat's udder before milking. *Right.* Udder after milking.

letters AR appear on the registration certificates of your doe's parents and grandparents it is reasonable to assume that she herself has promise of being a good milker.

Your neighbor may tell you of someone she knew whose goat gave six or seven quarts. Perhaps so, but people are funny about goats. Sometimes they tell tall stories, like the fisherman with his big catch. Six- and seven-quart milkers can be found, but not usually in a home dairy.

In purchasing a milking doe you may arrange to see her milked and know just what she is producing. If she is not milking you will have to trust to the integrity of the seller and judge from her past record or her breeding what she is likely to produce. Don't be disturbed and assume that you have been duped if, after you take her home, she drops in production. Goats are very sensitive to changes in environment, handling and feeding, and show it immediately in a lessened milk output. After she becomes accustomed to you, and to her new

Four does from one herd that won the Governor's Trophy for Best Eight Head, Toggenburg breed. Notice the breed characteristics: white strips from head to muzzle . . . white around tail and hocks.

home, she will give more milk, although it sometimes takes weeks before a doe is back to maximum production. Don't assume either that the doe will keep up a steady output for the entire period of lactation. She has a high peak, usually two or three months after kidding, and a low ebb as she nears the end of her lactation period. The figures above are averaged.

## Getting Your Goat.

FIRST you will wish to know what type of goat to get, where to get it, how much to pay for it, and so forth.

Before the war, goat-keeping had reached a degree when the bigger breeders considered that a goat giving anything less than five pints as her average was not worth keeping. Even so, a great many people with shorter purses found a two- or three-pint milker a good investment. And she is even more so in wartime.

It has been said that a small milker is as costly to keep as a good milker. This is not true. Normally, a two- or three-pinter will produce on pretty well what she can pick up for herself in summer. The production of yields higher than this entails the feeding of concentrates and this obviously puts up costs of production.

To economise on concentrates must now be every goat-keeper's aim, natural foods being used as much as possible. The wealthiest breeders even have had to modify their ideas in this respect.

Very true is it, however, that a scrub will not yield so well on the same rations as a well-bred goat of good milk strain. It is true, too, that a poor goat will cost *more* to keep than a good one, if by the former is meant poor in health and condition and upbringing.

My advice, then, is to get the best goat you can afford, and if you can afford only a two- or three-pinter, then, providing she is perfectly healthy and hardy, don't despise her, for as a result of a season or two's breeding you can produce from *her* even better and still better milkers.

A friend of mine started with a couple of scrub goats he bought for £1. He obtained an average of three pints between them. Feeding cost him roughly a shilling a week. He was well content with the saving in his milk bill—21 pints of milk weekly for 1/- which, bought at the door, would have cost him at least 5/-.

One of the goats I once bought myself was a half-pedigree, the daughter of a scrub female mated to a fully pedigree stud male. She cost £3/10/0. She kidded in April, and from then until September gave me an average of seven pints daily, then gradually declined through autumn and winter to not less than five pints. Her feeding cost me 3/- a week. The returns were, apart from some useful kids, and a daily output of splendid manure for the garden, 49 pints of milk a week—12/- worth if bought at the door at 3d. a pint. I, too, was well satisfied.

Another friend of mine paid ten guineas for a pedigree goat. He was unfortunate. Her average was no more than my non-pedigree's. She cost about a shilling a week more to feed as she had been brought up rather fastidiously. But wise mating, and the sale of good kids, soon put him on the right side.

The point is that there are good and bad milkers in both pedigree and non-pedigree goats. Pedigree or breed alone is insufficient recommendation for purchase.

A resumé of the various breeds will be helpful to those who know absolutely nothing about modern goats:

The Anglo-Nubian goat is the Jersey cow of the goat world, giving the highest butter-fat (cream) percentage of all goats, if not such a high yield as the Swiss types. A big animal, with a fine skin and glossy coat of varied colour, its very distinctive features are a Roman or camel-like nose and pendulous ears. All other breeds have prick ears. This breed is generally considered unsuitable for cold climates. While most affectionate and intelligent animals, they are noisy, which is a point to consider when thinking in terms of back gardens.

The British Alpine is a breed evolved by English breeders, a big animal, black with white or fawn markings, smart of appearance. A goat well known for its placid, quiet nature, and a heavy yielder of milk with averagely good butter-fats. Needs rather more exercise than smaller goats.

The Saanen is the white Swiss breed; a small, affectionate, heavy milking and long-lactation animal from which has been evolved the British Saanen. The latter, also pure white, is a larger, more heavily-built goat, a very heavy milker with excellent length of lactation. This breed has broduced two world goats' milk yield record holders.

The Toggenburg is a small, gentle and quiet Swiss breed, a good milker consistent in yield and excelling in long lactation. From it has been bred the British Toggenburg, which is larger and finer coated than the Swiss, with colour varying from the drab and white of the former to dark chocolate and white. The British type also has a slightly raised bridge to the nose; the Swiss type a straight or dished facial line.

Then there is what is known as the British breed. Usually of Swiss type, these are goats in which breeding, or parentage, colour, or " points " do not quite qualify them for registration in other breed sections. Excellent milkers and attractive animals, they are highly esteemed.

In all the above, the animals may be horned or hornless. Few pedigree horned goats are to be seen nowadays, however.

There are two other breeds indigenous to this country—the English (or Old English) and the Welsh. The former is small, short-legged, thick-set, horned, and very hardy. It is a great pity that little attempt has been made to improve yield with this breed for, while they are not heavy milkers, they are steady milkers, especially in winter, and butter-fats are excellent. The original Welsh is short-legged, small-bodied, long-haired and long-horned. The Improved Welsh Goat is now much in evidence in the Principality, of better size and type, greater milk yield, and shorter hair, thanks to careful breeding.

Except for the Anglo-Nubian, it is immaterial whether a goat has or not tassels, or fleshy appendages on the neck. They have no connection with milk yield.

There are cross-breds to be seen, of course, such as Nubian-Toggenburg, and half-pedigrees will be found answering closely to specified breeds. They are frequently advertised as, say, Saanen type, or Toggenburg type, or B.A. type.

At the present time the average householder's requirement will be for a moderate but steady yield of five pints, or so daily, for the lactation, from a healthy, hornless, attractive quiet animal capable of yielding on wartime rations. This average means that at the peak of lactation soon after kidding, the goat will give for some time seven to eight pints daily. Then is the time you will be rearing kids, making butter and cheese, and feeding surplus skim milk to your poultry, rabbits, or back-garden pigs.

A good half-pedigree goat will cost from £3 to £6 if purchased from a reliable source. A pedigree milker will cost considerably more, breeders roughly valuing their stock at £1 per lb. (nearly enough one pint) of milk yielded at full flush up to 10lbs.; 30s. per lb. for goats giving 10 to 15lb. daily. There is also reputation to pay for.

If a definite breed is desired, choose that, if possible, which is favoured in the district. It will be acclimatised, and there should be good males available for breeding.

Be careful where you buy. See the goat if you can, or purchase on approval only. Beware of the advertisement which offers gallon milkers from 30/-. Such do not exist. Choose a recognised breeder, or make sure you get a personal recommendation or unbiased outside opinion. You can obtain advice, and the addresses of genuine goat-keepers, quite freely.

Should you buy a milking goat, an in-kid goat, or a dry goat ? is a question you may raise. The in-kid goat is the best proposition. She has time to settle down with you before she comes into milk. Journeys and new quarters upset the yield of milkers; it may be difficult to bring back. A dry, unmated goat has nothing to compensate for her cost of keep, and a mate has to be found for her which may possibly incur trouble and expense. However, we cannot pick and choose in wartime, and as goats are so scarce many people will have to make the best of whatever is offered.

**Initial expense can be saved** by buying a kid or goatling. Here again the cost of rearing and keep before there are any returns must be considered. Fortunately, there are inexpensive and simple methods of rearing kids into good milkers.

A mated goatling will not be found the best "buy" for a beginner. As a first-kidder, she will have to be broken in to milk. Best for him to gain experience with a goat at her second or third kidding and already well accustomed to being milked.

A goat's best age is from four to six years, at her third or fourth kidding. Her yield then diminishes at each kidding, and usually drops to meagre proportions after nine years. The natural span of a goat's life is about 12 years. Several cases have been recorded, however, of goats living up to 18 years and yielding a pint or so daily.

It is now usual for well-bred goats to milk continuously for two or more years without an intervening kidding, the yield reducing in winter and picking up again in spring. Such goats are very useful for those who don't want the bother of dealing with kids every year.

One goat may be found sufficient for a time, but you will surely experience the desire for two or more later on. Make your plans accordingly. Arrange for one, if you wish, but plan for two. Don't, for instance, get a house which will only hold one. There is very little extra expense in a two-goat house.

Two goats will ensure a satisfactory milk supply all the year, mated at different seasons. Three goats will be more than adequate if the system is adopted of mating one in September to kid in February; another in January to kid in June; and the third left unmated to milk through, the order being reversed the following year.

In any case, goats are companionable to extreme, keeping happier and yielding better in company. A lone goat, especially if bought from a herd, will fret, or will attach its affections to its owner, sometimes to embarrassing extent, following him or her around like a faithful dog—which isn't always convenient !

Finally, it should be pointed out that, unless you are buying a youngster, the goat you obtain should be used to the conditions it will meet with you. An old goat accustomed to free range will not do well confined.

# CHOOSING A GOAT

Health is more important than anything else. Unless a nanny has *stamina* (good health and a well-grown body) she will never be a satisfactory producer of milk and breeder of kids, and she will not have stamina if, as a kid, she was under-fed or over-fed, neglected or coddled. Furthermore, if her parents were not healthy she herself may be a weakling. And the health of *her* parents depends in turn upon the health of *their* parents. Therefore, whenever possible, a goat should be bought from a herd about which something is known, and from a breeder who understands the importance of health.

**The Signs of Health.**

1. The coat of a healthy animal is glossy and smooth. This gives it *bloom*. If the animal is unhealthy, however, the coat loses its glossiness and is said to be *staring*.

2. A *bright and noticing eye* is a sign of good health. The animal that is sick has a dull eye with a distressed, listless or alarmed expression.

3. A healthy animal carries itself well, is light on its feet, and its eyes, nose and ears show that it is interested in its surroundings. This is what is meant by an *alert manner*. A sick animal may stand about and mope, with its back arched and its ribs drawn up (*tucked up*).

4. A healthy animal, properly fed, is always eager for its food. It has a *good appetite* that can be satisfied.

There are many signs of ill-health, and these are described on page 41.

**Milking Qualities.** If our goat is healthy we can next consider her as a producer of milk. In this, also, her ancestry is important, for the capacity to produce milk is usually inherited, and a goat descended from a long line of good milkers is *more likely* to milk well than one whose female ancestors were but poor milkers. Her milking qualities depend, however, as much upon the male as upon the female parent. This may sound strange, but it is often proved that the daughters of some males are better milkers than the daughters of others, even when these males are mated to the same nannies.

17

6.  The equipment needed for tethering is a broad, smooth, strong, leather collar, a stout chain, and a steel pin about 18 inches long.   The chain must not be very heavy, and must be fitted with a swivel at each end so that it does not get twisted as the goat moves round. How long would you judge this chain to be ?  What are the drawbacks to a very long chain ? The pin in the picture is a very long one, but has been driven in quite securely.

7.  An intensive house, with a corner of the exercising yard on the right.  Corrugated iron is a good conductor of heat and therefore makes a shed hot in summer and cold in winter.   Matters can be much improved by packing the underside of the roof with straw.

A goat that gives 6 to 8 pints a day when her yield is highest must be called a good milker. She will probably give 150 to 200 gallons in the year. One who at no time gives more than 4 or 5 pints a day is a moderate milker. Her annual yield may be 80 to 120 gallons. A few goats have given as much as 400 or 500 gallons in a year. The standard varies somewhat with the different breeds. See page 16.

The maximum yield usually comes in the second or third month after kidding, though the growth of fresh herbage in spring influences this. From the peak onwards the daily yield becomes less and less until the goat is dry. See page 34.

A goat that *keeps up her yield* may be giving a pint a day, or even more, a month before kidding again. Others may be dry 3 months or more before kidding.

Most goats milk best with their third kid, but there are many exceptions to this rule.

A quiet, good-natured disposition is desirable in a milking animal, and when buying a goat in milk it is as well to see her milked before completing the bargain. To be certain of her yield it is necessary to witness two consecutive milkings.

Some goats are more economical than others in turning their food into milk. A goat that gives a moderate quantity of milk on the common, inexpensive foods, is probably more economical than a very high yielding animal that needs large amounts of expensive *concentrates*. See page 23.

The *quality* of the milk is also important. This matter is dealt with in Chapter 5, and all that need be said at the moment is that quality depends largely upon the amounts of *fat* and *other solids* in the milk.

Knowledge of a goat's ancestry is not always available. Nor is it always reliable, for the result of the very best breeding is not always a good goat. We must, therefore, learn to judge a milking goat by the shape and proportions of her body, as far as such judgement is possible. It is easiest to do this when the goat is in milk, and Pictures Nos. 13 to 16 illustrate the principles.

**Breeding Capacity.** A female goat does not normally give milk until she has been mated with a male and, as a result, has given birth to a kid.* Therefore, for a goat to produce her maximum yield of milk she must be a regular

* It is, however, not uncommon to find goatlings that milk before being mated. In such cases the udder has to be emptied as often as it feels full. It is suggested that goats that possess this rather surprising character are never poor milkers, though all good milkers certainly do not possess it. Unmated milkers frequently go dry in the winter.

8. The goat house need be neither expensive nor elaborate.  This picture shows a range of pigsties adapted to house goats.  The verandah gives added shelter, and the concrete path means greater comfort for both stockman and goats.  How is the hay being fed? What breeds of goats can be identified in this picture?  To answer this, see pages 14 and 17.

9. A useful shelter for a tethered goat.  The need for such a shelter is explained on page 3.  What are the advantages and disadvantages of its being so very light?

breeder. The general rule is to breed from the females once a year, mating them 7 or 8 months after the previous kidding. Occasionally, however, a goat keeper may decide not to mate a particular goat one year, and such a goat, not being in kid, may continue to milk for 2 years or even longer.

**Pedigree.** Any animal whose ancestry is known is really a *pedigree* animal, but the word is usually reserved for those animals *registered in* (entered in) the Herd Books of the breed societies. Entry into the herd book depends entirely upon ancestry. (See the regulations of the British Goat Society.)

The phrase, *pedigree stock*, suggests, to some people, animals that are delicate and in-bred. Such animals are the exception and not the rule.

**Pure-Breeds and Cross-Breeds.** The majority of goats in this country belong to breeds that have been *evolved* (developed) by crossing our native goats with imported goats. This cross-breeding has been carried out for a great many years, with the result that these " new " breeds are now firmly established and the goats that belong to them generally breed *true to type*. This means that the young resemble their parents and have the characteristics of their breed.

**Native Breeds.** *The Old English Goat* was, until recently, the one distinct native breed. See Picture No. 12. There are very few pure-bred animals left, and the proportion of English blood in the goats of our other breeds is getting less and less as time goes on. Nevertheless, as this breed was *indigenous* (native) and therefore well suited to our climate, it helped to produce some remarkably hardy goats.

*The Welsh Goat,* as a pure-bred, is now extinct. There are in Wales, however, a number of goats that are the product of a cross between the native female and males of Swiss blood (see below). This cross-breeding has brought about a great improvement in the milk yields, while the hardiness that is always found in goats of an indigenous breed when kept in their native lands, remains in many of these Welsh-Swiss crosses.

**Imported Breeds.** Four different types of goat have, in the past, been imported into this country.

(1) *The Oriental (Eastern) type,* known as Nubian, with arched face and drooping ears. There are now no pure-bred oriental goats in this country.

# PLAN OF SMALL GOAT HOUSE

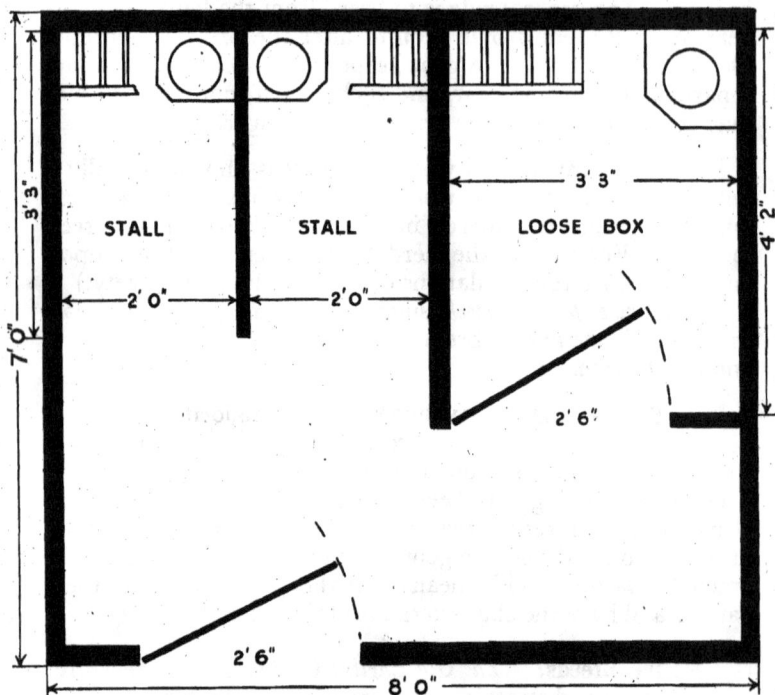

10. In this plan the hay rack is shown broken in the middle so that the board that holds the two feeding buckets is visible. The spaces between the bars should be about 2 inches wide. Hay nets can be used instead of a rack. See Picture 19. The boards that hold the feeding buckets are about 2 feet above the ground. Height of loose box walls : 4 feet. Height of stall partition : 4 feet at the head sloping to 3 at the back. Partition and walls extend to the ground. The measurements of the loose-box should, if possible, be a little greater than those in this plan, for these are the minimum dimensions.

Floor : concrete, sloping to gulley and drain. The laying of a concrete floor is described in " Pig Keeping ", Booklet No. 4 in this series. Racks, similar to the one shown in Picture No. 5, can be made to fit into the stalls and into a corner of the loose box. They save bedding, but are unnecessary if there is plenty of cheap material to use for litter.

The goats are tied up in the stalls by means of collars and chains. Each chain is free to slide up and down a vertical iron bar fixed to the stall partition.

Any shelves put up to hold equipment must be out of reach of the goats. They are very inquisitive animals. They are also very strong, so all the partitions and doors must be stoutly built. The wood should be creosoted in order to preserve it and to discourage the goats from chewing it. To what scale is this plan drawn ?

Many goat keepers have to house their goats in existing buildings, and they are therefore obliged to make do with what they have. When a new building is to be erected, however, more perfect arrangements can be made, and it always pays to give long and careful thought to the lay-out. A rather better plan than the above simple one might then be designed. Plans and advice can be obtained from the British Goat Society, as a reference to page 47 of this booklet will show.

(2) *The Toggenburg from Switzerland.* There are a few pure-bred herds of these goats descended direct from the imported animals without crossing with other goats.

(3) *The Saanen from Switzerland.* There are a limited number of pure-bred herds in this country.

(4) Other Swiss types, found in many varieties throughout the Alps. There is no pure-bred stock in this country, the imported goats having been used only for crossing.

## Breeds Developed in Great Britain.

(1) *The British.* When the breeding of goats was first developed in this country the main object of the breeders was to 'improve the size, stamina and milk yields of their animals. The imported breeds were freely crossed with one another and with our native stock in order to achieve this object, and colour, markings and the smaller details of appearance were largely ignored. The resulting animals were originally known as Anglo-Nubian-Swiss, but are now called British. Although, as might be expected, they vary greatly in type and colour, their general appearance is similar to that of the Swiss breeds. Thus, they are sometimes described as " of Saanen type ", or " of Toggenburg type ".

(2) *The Anglo-Nubian.* Formed by crossing imported goats of the oriental type with our native goats. See Picture No. 11.

(3) *The British Toggenburg.* Formed by crossing the Toggenburg with the other Swiss, the British and native goats, and carrying enough Toggenburg blood to retain the colour and the general character of that breed. The colour varies from dark chocolate to pale fawn, and there are white markings on the hind-quarters and on the sides of the face.

(4) *The British Saanen.* Formed in the same way as the British Toggenburg, but substituting Saanen for Toggenburg as the basis. See Picture No. 13.

(5) *The British Alpine.* A breed based upon the Swiss breeds as well as upon British and native stock. The distinctive colouring has been fixed by breeding from selected animals. See Picture No. 14.

A goat keeper who is concerned mainly with the production of milk need not necessarily invest in pedigree, registered

11. Anglo-Nubian female kids. Goats of this breed have pendulous (drooping) ears and a Roman nose (arched or convex profile). The coat is short and may be of many colours: black and tan, mottled, or the same all over. Goats of this breed are, perhaps, slightly less hardy than those of other breeds, but strains and herds vary in this respect.

12. Female goat of the Old English breed. The Old English goat was a small, compact animal, short on the leg. The coat was rough and thick. Goats of both sexes were horned and bearded. The face was dished (concave), and the ears were prick. Tassels were absent.

goats. If a mongrel goat of a good milking, hardy *strain* (see below) is mated to a pedigree, registered male of good milking ancestry, there is every chance of producing some female kids that will become good milkers in their turn. If this procedure is followed for a few generations we may well find ourselves with some very useful milkers who will qualify for admission into a herd book.\* This is called *grading-up*.

To make this improvement possible pure-bred males are needed, so the registered, pedigree herds must be kept in order to improve and maintain the quality of our stock. Continual and haphazard breeding from cross-breds produces, sooner or later, inferior animals.

Milking competitions and the efforts of breeders have succeeded, during the last twenty years, in raising the general level of milk yields by over 50 per cent, and, on the whole, this has been done without any corresponding loss of stamina.

**Choice of a Breed.** 1. *Strain* is more important than breed, for there are good and bad goats in every breed. The term *strain* corresponds roughly to the word *family* as applied to human beings. All the goats of one strain are related to one another, more or less closely, either directly or through an ancestor common to them all. They will frequently, but by no means always, bear a close resemblance to one another. Some strains are more hardy than others of the same breed, some contain many excellent milkers, some are noted for the richness of their milk, and so on. It is, therefore, very important to choose a goat of a good strain, but this is only possible if something is known about the milking and breeding capabilities of the parents and other relations.

2. Goats with a good proportion of native blood in them are particularly well suited to exposed situations. Goats of the English and British breeds, many crosses between these and the Swiss breeds, and the Welsh-Swiss crosses are generally very hardy.

3. On the average, the British and Swiss breeds contain the heaviest milkers, and the Anglo-Nubians come next. There are, however, such great variations between individuals that strain must be considered first. The Anglo-Nubians are noted for the richness of their milk in butter-fat.

4. Most people have a fondness for goats of one particular breed, either because they are attracted by their appearance

---

\* Unless a goat qualifies for admission to the appropriate breed section of the herd book, it should not be described as belonging to that breed.

13. The body of a dairy goat should be deep, but deeper behind than in front. This makes it wedge-shaped when viewed from the side. The back should be level, though this does not mean a perfectly straight line. The British Saanen above has a level back. The forelegs should be straight, and all four legs look strong without being coarse and clumsy ; this is what is meant by having good bone. The skin should be loose and supple.

14. British Alpine female. Black with white markings. Short coated. Usually hornless.

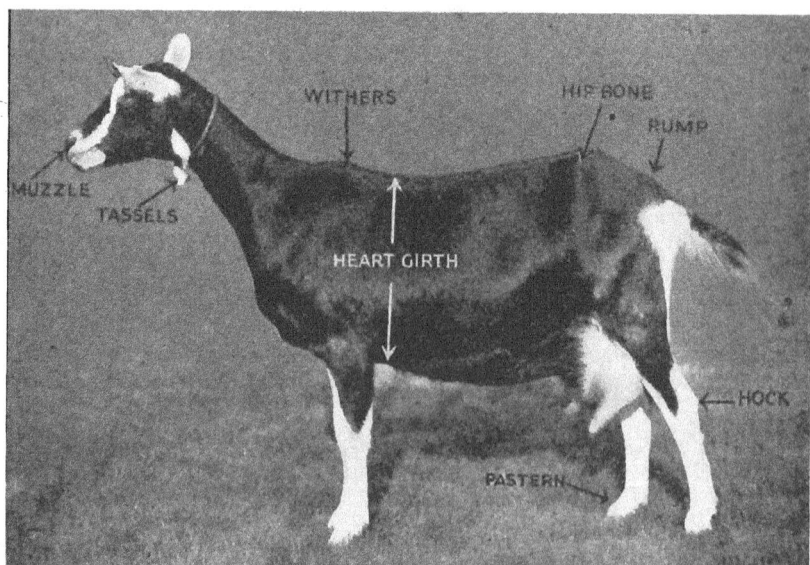

or because they already know something about them. This is a very good reason for choosing them, but it must not be allowed to over-rule more important considerations.

5. Sometimes a friend may offer us a goat at a low price, but such an offer should be carefully considered in the light of our knowledge before it is accepted.

6. If one particular breed is popular in a district there may be a good reason for this, and it might be as well to follow the local example. The presence of a good male in the neighbourhood may make us decide to keep goats of the same breed so that the herd can be kept pure without our having ourselves to keep a male. There is, therefore, no one answer to the question " Which is the best breed ? "

**Age at which to Buy.** The age of a goat can be told approximately by an examination of her front teeth (*incisors*). See " The Book of the Goat ", by H. S. Holmes Pegler. If we buy a goat a few weeks after *kidding* (giving birth to a kid) we shall have a goat in full milk. This is generally the most expensive way of starting, but we shall get a return for our money at once.

If we buy a goat that has been milking for some time and is in kid again, we may have a goat that is giving a little milk. When she kids she comes into full milk again. A goat in kid may cost as much, or nearly as much as a goat in full milk.

If we buy a *goatling* (a young goat between 12 and 24 months old) we shall have to wait longer before we get any milk than if we buy an older goat *in profit*. This goatling may have been mated before we buy her, or she may be too young for this. In the latter case we shall *run her on* (keep and feed her) until she is ready for mating.

The cheapest way of all is to buy a female *kid* (a goat less than 12 months old) and rear her. We shall, however, have to wait a long time before we get a return, and the kid will need a good deal of food, including milk or a milk substitute.

Goats like company, and for this reason it is best to keep more than one, if possible. If two goats are to be bought they can conveniently be of different ages. For example : a goat in milk and a goatling in kid.

We *must* learn to milk before the goats arrive. This is best carried out on a goat (or a cow) that is going dry, because an animal in full milk can easily be damaged by an inexperienced milker. Milking cannot be learnt in a day. See page 37.

*Prices.* The price of a milking goat varies with its capacity to give milk. Previous to 1939 a rough guide to price was to reckon £1 for every pint of milk given at the time of maximum yield. Thus, a goat giving 8 pints of milk a day was worth about £8. If she was a pedigree goat she might have been worth half as much again, i.e. about £12. The war, however, produced a keener demand for goats, and their price rose in consequence. To-day (1948) prices are about 75 per cent more than those given above. Thus, a non-pedigree goat giving 8 pints a day at her maximum is worth about £14, and a pedigree goat with the same yield will cost about £21. Other considerations also affect the price.

**Billies.** The *points* (bodily characters) of a well-bred male goat are shown in Picture No. 24. These, if possible, should be studied in conjunction with his pedigree. Indeed, most people place pedigree first and points second in the choice of a male. The choice is most important, for he may, as explained on page 8, transmit to his daughters the milking qualities of his dam, grand-dam, and other female relations in the direct line. His daughters, moreover, are more numerous than the daughters of a female goat, so his influence upon the herd is even greater than theirs.

An unregistered male with bad points and *throwing* (being the sire of) bad stock, is called a *scrub* goat. More often than not he is a *mongrel* (of mixed and unknown ancestry). Such a goat should not, of course, be used for breeding. On the other hand, a billy whose nanny kids develop into milkers as good as, or better than their dams, is called a *proven sire*. His worth has been proved, and he is a very valuable animal indeed. He may or may not be a pure-bred and pedigree animal. A male cannot be proven until his daughters come into milk, which is, at the earliest, when he is 3 to 4 years old.

It is, of course, an expense to keep a male goat, and if only one or two goats are kept and a neighbour has a good male, it is unnecessary for us to keep one ourselves. If, however, we own a number of goats it is advisable to have a male. The nannies can then be mated on the farm, thus saving the time and expense involved in taking them, perhaps long distances, to a farm where a male is kept. It is also frequently found that where a male is kept on the place fewer of the nannies *return*. See page 28.

A well-grown male is usually fit for *service* (mating) at the age of 6 months, but he must not be used too frequently in his first season.

**The Stud Goat Scheme.** This is organized by the Ministry of Agriculture and the British Goat Society in an attempt to improve the milking qualities of goats kept by smallholders and cottagers by making available at a fee not exceeding 7s. 6d. the services of well-bred male goats. It has proved very successful. Further particulars of the scheme can be obtained from the British Goat Society. See page 47.

Male goats kept for breeding are called *stud goats*.

**Horns.** Some goats grow horns naturally, and others, equally naturally, have none.

Breeders have tried so hard to *breed out horns*, by continually breeding from hornless animals, that nowadays a large proportion of pure-bred and cross-bred goats are naturally hornless. Some kids, however, even though their parents are hornless, show signs of developing horns. These animals can be *disbudded*. (See page 46.) Most goat keepers prefer hornless animals, for accidents are then less likely to happen. Nor is a hornless goat so likely to get caught in, or do damage to the fittings. On the other hand, horns are a useful protection against dogs for a goat tethered in the open, and they also form a useful handle if a goat gets saucy. In the Old English breed horns were one of the breed characteristics, but in the Anglo-Nubian and the Swiss and allied breeds, horns are discouraged.

**Tassels.** These are sometimes called *doddles*, and American goat keepers call them *wattles*. They are shown in Picture No. 14. Very many, but not all, goats carry them. Numerous suggestions are made regarding their origin, but nothing definite is known.

**Beards.** Nearly all male goats grow beards, and many females do so too. The size of the beard varies considerably between one individual and another, but as a goat grows older so its beard becomes longer.

**Coats.** English and Welsh goats, and those crosses in which this blood predominates, carry heavier coats, as a rule, than the Swiss and Anglo-Nubian breeds, although some strains of pure Toggenburgs and the males of the Saanen breed are often long coated. The thickness of the coat, however, also depends upon the conditions under which the goat is kept. A heavier coat is grown when the animal is kept under open-air conditions than when it is housed most of the time, is groomed regularly, and in cold weather is frequently *rugged up*. A sleek, light coat is preferred when exhibiting a goat at the shows, but a heavy coat is a blessing to a goat that is tethered or kept in a simple type of house.

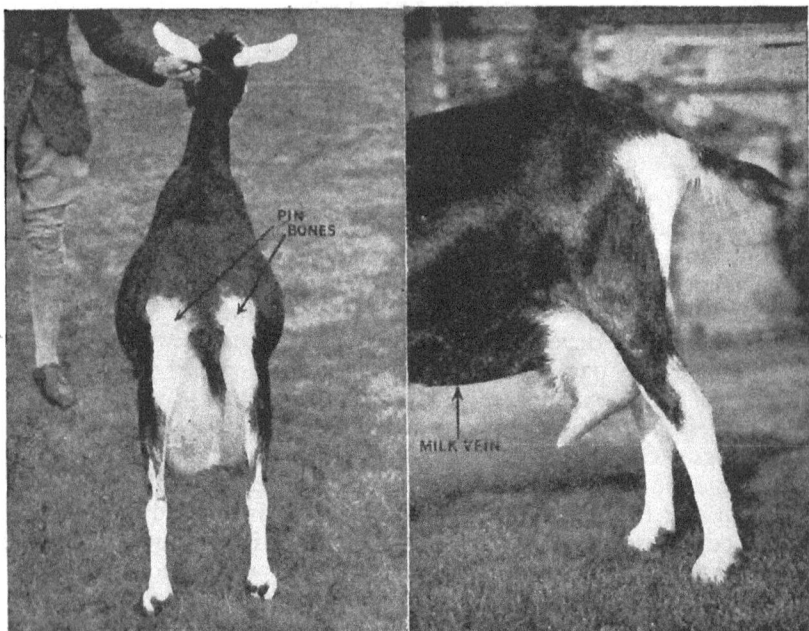

15. When choosing a goat by her appearance we should select one that is wide across the hips, wide between the pin bones, and wide between the hocks. The ribs should be well sprung (arched). The withers should be fine, i.e. narrow above the shoulders. The fine withers and the wide hips make the goat appear wedge-shaped when viewed from above.

16. The rump should be long and slope gently down towards the tail. The perfect udder is similar in shape to a globe, and the two halves should be as nearly alike as possible. It should be carried well forward and be attached to the body over a wide area so that it does not hang from a neck. The milk veins should be well developed, i.e. prominent and long.

17. This shows the facial markings of the British Alpine. The Toggenburg has somewhat similar white markings on the face. Are these the faces of goats or goatlings?

# TOGGENBURG BREED

This breed is one of the leading breeds of Switzerland and is named from the Toggenburg Valley where it has been bred for many years. The Swiss seem to rely to some extent upon the high mountains inclosing each valley or canton for protection from invasion by man or beast and it usually serves that purpose though it is not impenetrable. There are many such more or less secure valleys and it is said some sixteen varieties of goats exist in Switzerland, though close comparison suggests in some instances they are mere branches of an original line as indicated by type, size, color and general characteristics and this applies to each of the two outstanding breeds originally imported from there to the United States.

The present Advanced Registry record held by this breed is by Crystal Helen 48693 Advanced Registry 350 of 3726.2 lbs. milk during ten months official test and the high Butter Fat Test Lahoma Pat. 60414 Advanced registry 474, a ½ grade Toggenburg produced 133.231 lbs. butter fat during nine months 21 days official test.

TOGGENBURG DOE

The color must be brown though permissibly ranging from a light to dark shade with regulation white stripe down each side of the face, white edges of ears, white each side of tail and all four legs white below the knee and hock joints. Hornlessness predominates but by no means is it complete.

TOGGENBURG BUCK

There was a time when the appendages known as wattles were regarded as a necessity in the estimation of extremists though that idea has faded.

# TOGGENBURG STANDARD

At present the Committee on Standard submits the following for All Breeds though this to date has not been officially adopted nor can it be until the next annual meeting of the Record Association though it does give a comprehensive idea as to points to be observed.

## *Descriptive Standard for Toggenburg Doe*

### HEAD 8 POINTS

Head of medium size, fine in contour, feminine; facial lines straight or dished; eyes large, clear, bright; broad between eyes; ears erect or pointing forward ......... Score    3 points

MUZZLE broad; strong nostrils, lips, jaws ..................... Score    3 points

POLL—naturally hornless credit 2 points; neatly disbudded without scurs, credit 1½ points; horns or large prominent sawed-off stubs, discredit 2 points ......... Score 2 points

## NECK 3 POINTS

NECK long and slender; with or without wattles, with no dewlap, smoothly joined to head and shoulders; wattles, if present, should be evenly hung ............ Score 3 points

SHOULDERS slightly lower than hips; light and smoothly blended into body; withers thin and sharp; wide between shoulders; chest deep and wide between and back of front legs; forelegs straight and strong Score 6 points

BODY long with well sprung ribs and ample heart girth; deep and well developed abdomen firmly held up; back straight and strong with broad long rump, only slightly sloping; hip bones wide apart; thurls preferably somewhat wider apart than hip bones; vertebrae open, with prominent spinal processes ...... Score 10 points

## BODY 22 POINTS

HIND-LEGS clean, strong, straight, proportionate as to size, without tendency to being cow-hocked; hind quarters deep, somewhat concave behind, allowing plenty of width for udder, upright pasterns, feet true, pointing forward ......... Score   6 points

UDDER globular, capacious attached over a large area; free from fleshiness, flexible and of fine texture; attached well forward at front, with no cavity between the two halves; at rear well rounded, attached high and firmly, with no signs of being pendulous; halves even, with no indentation between halves and so smoothly joined as to present a blended rounded contour at rear ................... Score 25 points

## MAMMARY SYSTEM 37 POINTS

TEATS symmetrical, pointing downwards and slightly forward; moderately long and of a size that can be comfortably held in the hand for milking; teats should be squarely set, wide apart, distinct from udder, not blending into udder; good orifice for milking strong stream of milk .................. Score   8 points

MILK VEINS large, long, prominent, tortuous, with lateral branches . Score   4 points

## SIZE AND GENERAL APPEARANCE 30 POINTS

SIZE—Mature does; height at withers 27 inches and up; weight 125 lbs. and up ................. Score  10 points

GENERAL APPEARANCE—Active, vigorous, true to type, hair soft, fine and glossy; skin loose and pliable. Score  15 points

COLOR AND MARKINGS—solid color, varying from light fawn to dark chocolate, with no preference for any shade; marked distinctly with white as follows: white ears, preferably with a dark spot in the middle; two white stripes down the face from above each eye to the muzzle; white muzzle, legs white on the inner side

and entirely white from the knees
and hocks to the hoofs; a white triangle on either side of the tail;
white spot at root of wattle, or in
that area, if no wattles present. Cut
for extra body spots, but not to such
an extent as to overcome major
points of excellence. Black coat disqualifies ................ Score   5 points

## Descriptive Standard for Toggenburg Buck

### HEAD 17 POINTS

Head, masculine and virile, clean in
    outline, strong but not coarse; facial
    lines straight or dished; eyes large,
    clear, bright; broad between eyes;
    ears erect or pointing forward Score   6 points

MUZZLE broad; strong nostrils, lips,
    jaws ................... Score   6 points

POLL—naturally hornless credit 5
    points; neatly disbudded without
    scurs, credit 3 points; horns or large
    prominent sawed-off stubs, discredit
    5 points ............... Score   5 points

## NECK 5 POINTS

NECK strong and heavy, with or without dewlap; with or without wattles; wattles if present, should be evenly hung .............. Score 5 points

## BODY 38 POINTS

SHOULDERS may be even with or slightly higher than hips; strong and powerful, blending smoothly into body; withers thin and sharp; wide between shoulders; chest deep and wide between and back of forelegs; forelegs straight and strong with good bone development .... Score 10 points

BODY long with well sprung ribs and ample heart-girth; deep and well developed abdomen firmly held up; back and rump strong, with rump only slightly sloping; hips wide apart ................... Score 18 points

HIND-LEGS straight, strong, facing forward; wide between hocks; upright pasterns, feet true ........ Score 10 points

## ORGANS OF REPRODUCTION 10 POINTS

ORGANS OF REPRODUCTION—well developed, showing breeding ability. Correct number rudimentary teats, namely two ............. Score 10 points

## SIZE AND GENERAL APPEARANCE 30 POINTS

SIZE—mature bucks; height at withers 33 inches and up; weight 160 lbs. and up ................. Score 10 points

GENERAL APPEARANCE—active, vigorous, rugged but not course; coat smooth and glossy hair short or medium length preferred; skin thin and loose .............. Score 15 points

COLOR AND MARKINGS—same as for does, except that the two white lines on the face often fade out on a mature buck, therefore no cut is recommended on this. Large white spots on side of body disqualify

Score 5 points

NOTE—Buck kids may be judged by the same standard as MATURE BUCKS.

## Standard for Toggenburg Doe Kids, and Yearling Does Not in Milk

KIDS and YEARLINGS should be judged on the basis of the promise which they show of developing into mature does of high quality as described in MATURE DOES' STANDARD. This does not mean, however, that a kid with a precocious udder should be placed ahead of a kid with normal udder development. If a kid has the development normal to her age, she should receive full credit in her score on udder development. Similarly when milking yearlings are shown in the same class with yearlings not in milk, no advantage should be given the milking yearling for her superior udder development as compared with the yearling not in milk.

## Disqualifications

Unsoundness of any kind, disease or deformity disqualify a goat in the show-ring. Any artificial means of removing or remedying any defects by cutting or filling under the skin disqualifies, with the exception of dehorning, disbudding, clipping and trimming the hair. Hermaphrodites and castrated goats are ineligible to show. Dying the coat is considered as fraud and deception and **disquali-**fies the animal.

The preceding constitutes the Majority Report on the Toggenburg Descriptive Standard.

The Minority Report concurs with the Majority, except on the following points:

> Under Toggenburg Does, COLOR: **Large** white spots on side of body, disqualify.
>
> Under Toggenburg Kids and Yearlings: Milking yearlings should be given preference due to the visible proof of their ability. Preferably should be shown in a separate class.

## SAANEN BREED

This breed is somewhat larger, is or should be clear white, though slight tinge of creaminess is permissible due to effects of sun, etc., but should not be cultivated, in fact, is being eliminated by rigid culling of off colored bucks. However, a choice heavy producing doe even slightly off color, should be preserved and mated to a clear white buck whose lineage is free from defect for at least two or more generations.

The Saanen is often spoken of as the Holstein, since she has led the long line of advancement in attaining world records with but few exceptions. Her record for production has made her a favorite with some for dairy purposes. For type, conformation, etc., the illustration is suggestive. Some critics say the Saanen is hard to keep clean but that is a mistake for any goat is extremely fastidious as to where she will walk, stand or lie down, and if her color or lack of color should prove to be more susceptible to dust, soot, etc., then that may be an indirect advantage, since it calls for extra grooming, for all goats of any breed worth keeping at all, are worth keeping well.

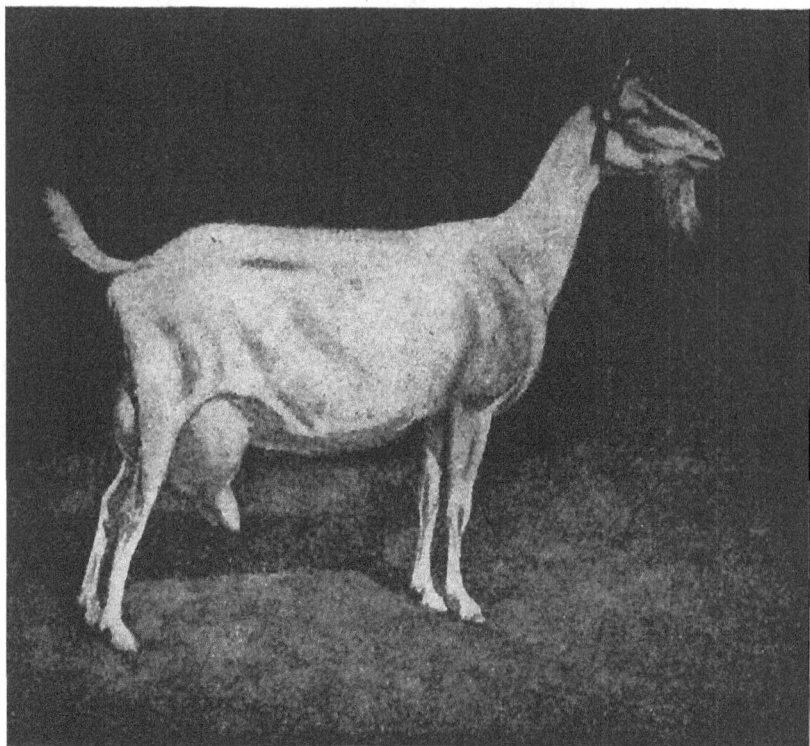

SAANEN DOE

The present Advanced Registry record held by this breed is Three Oaks Blossom's Marmaine 34288 A.R. 146. produced 4161.7 lbs. milk during nine months 8 days official test and the high butter fat test Lila of Ontario 57884 A.R. 469 Produced 142.715 butter fat during ten months official test.

SAANEN BUCK

# SAANEN STANDARD

The Committee on Standards for Saanen recommends the following but as yet has not been officially adopted nor can it be until the next annual meeting of the Record Association though it does give a comprehensive idea as to points to be observed.

## *Descriptive Standard for Saanen Doe*

### HEAD 8 POINTS

Head of medium size, fine in contour, feminine; facial lines straight or dished; eyes large, clear, bright; broad between eyes; ears erect or pointing forward .......... Score   3 points

MUZZLE broad; strong nostrils, lips, jaws ..................... Score   3 points

POLL—naturally hornless credit 2 points; neatly disbudded without scurs, credit 1½ points; horns or

large prominent sawed-off stubs, dis-
credit 2 points . . . . . . . . . . . . Score   2  points

## NECK 3 POINTS

NECK long and slender; with or without
wattles, with no dewlap; smoothly
joined to head and shoulders; wattles, if present, should be evenly
hung . . . . . . . . . . . . . . . . . . Score   3  points   .

## BODY 22 POINTS

SHOULDERS level with hips; light and
smoothly blended into body; withers thin and sharp; wide between
shoulders; chest deep and wide between and back of front legs; fore-
legs straight and strong . . . . . Score   6  points

BODY long with well sprung ribs and
ample heart-girth; deep and well developed abdomen firmly held up;
back straight and strong with broad
long rump, only slightly sloping;
hip bones wide apart; the thurls
preferably somewhat wider apart
than hip bones; vertebrae open,

with prominent spinal processes

Score 10 points

HIND-LEGS clean, strong, straight, proportionate as to size, without tendency to being cow-hocked; hind
quarters deep, somewhat concave
behind, allowing plenty of width for
udder, upright pasterns, feet true,
pointing forward ......... Score 6 points

## MAMMARY SYSTEM 37 POINTS

UDDER globular, capacious attached
over a large area; free from fleshiness, flexible and of fine texture; attached well forward, at front, with
no cavity between the two halves; at
rear, well rounded, attached high
and firmly, with no sign of being
pendulous; halves even, with no indentation between halves and so
smoothly joined as to present a
blended rounded contour at rear

Score 25 points

TEATS symmetrical, pointing downwards and slightly forward; moderately long and of a size that can be

comfortably held in the hand for milking; teats should be squarely set, wide apart, distinct from udder, not blending into udder; good orifice for milking strong stream of milk .................... Score **8** points

MILK VEINS large, long, prominent, tortuous, with lateral branches . Score **4** points

## SIZE AND GENERAL APPEARANCE 30 POINTS

SIZE—mature does; height at withers 29 inches and up; weight 130 lbs. and up ...................... Score **10** points

GENERAL APPEARANCE—active, vigorous, true to type, hair soft, fine and glossy; skin loose and pliable. Score **15** points

COLOR—white or light cream color, White preferred. Black spots allowed on skin, but not in coat. Score **5** points

### *Descriptive Standard for Saanen Buck*

## HEAD 17 POINTS

Head, masculine and virile, clean in outline, strong but not coarse; facial lines straight or dished; eyes large, clear, bright; broad between eyes; ears erect or pointing forward.

Score 6 points

MUZZLE broad; strong nostrils, lips, jaws .................. Score 6 points

POLL—naturally hornless credit 5 points; neatly disbudded without scurs, credit 3 points; horns or large prominent sawed-off stubs, discredit 5 points ................ Score 5 points

## NECK 5 POINTS

NECK strong and heavy, with or without dewlap; with or without wattles; wattles if present should be evenly hung ............. Score 5 points

## BODY 38 POINTS

SHOULDERS may be even with or slightly higher than hips; strong and powerful, blending smoothly into body; withers thin and sharp; wide between shoulders; chest deep and wide between and back of forelegs;

forelegs straight and strong with good bone development .... Score 10 points

Body long with well sprung ribs and ample heart-girth; deep and well developed abdomen firmly held up; back and rump strong, with rump only slightly sloping; hips wide apart .................... Score 18 points

Hind-Legs straight, strong, facing forward; wide between hocks; upright pasterns, feet true ........ Score 10 points

## ORGANS OF REPRODUCTION 10 POINTS

Organs of Reproduction—well developed, showing breeding ability. Correct number rudimentary teats, namely two ............. Score 10 points

Size—mature bucks; Height at withers 35 inches and up; weight 185 lbs. and up ................. Score 10 points

## SIZE AND GENERAL APPEARANCE 30 POINTS

General Appearance—Active, vigorous, rugged but not coarse; coat smooth and glossy; hair of short or

medium length preferred; skin thin
and loose ............... Score 15 points

COLOR AND MARKINGS—white or light
cream color. White preferred. Black
spots allowed on skin, but not on
coat .................... Score 5 points

NOTE.—Buck kids may be judged by the same standard as mature bucks.

## Standard for Saanen Doe Kids, and Yearling Does Not in Milk

KIDS and YEARLINGS should be judged on the basis of the promise which they show of developing into mature does of high quality as described in MATURE DOES' STANDARD. This does not mean, however, that a kid with a precocious udder should be placed ahead of a kid with normal udder development. If a kid has the development normal to her age, she should receive full credit in her score on udder development. Similarly when milking yearlings are shown in the same class with yearlings not in milk, no advantage should be given the milking yearling for her superior udder development as compared with the yearling not in milk.

## Disqualifications

Unsoundness of any kind, disease or deformity disqualify a goat in the show-ring. Any artificial means of removing or remedying any defects by cutting or filling under the skin disqualifies, with the exception of dehorning, disbudding, clipping and trimming the hair. Hermaphrodites and castrated goats are ineligible to show. Dying the coat is considered as fraud and deception and disqualifies the animal. A goat that does not show the characteristics of her breed is disqualified to place in the breed classes.

The preceding constitute the Majority Report on the Saanens Descriptive Standard.

The minority Report concurs with the Majority, except in the following particulars:

Under Kids and Yearlings, add: Milking yearlings should be given preference due to their visible proof of their ability. Preferably should be shown in separate class.

# NUBIAN BREED

Here is what might be termed the Bible goat, as the goat referred to in Biblical times, her outstanding characteristics include the convex face and pendulous ear, often reaching a point one to two inches below the chin. The color may range from black to white in any shade or degree even spotted and seldom is one found with any one color predominating completely. Her milk is extremely high in butter fat, which helps to balance the greater value of some others and yet she can and does produce as much as one should reasonably expect from an animal the size of our improved milk goats, for they all are veritable machines manufacturing feed into milk through which there is no one feed superior from any stand point of reason. The high milk test and high butter fat test for this breed is held by Creamy's First 46647 A.R. 354. Produced 1945.8 lbs. milk and 103.046 lbs butter fat during 10 months official test.

This breed possibly has a shade of right for the

claim of being more generally hornless than any other and her ardent admirers even go to the extreme of claiming the Nubian Buck being free from odors. Well, that may apply to any buck well groomed and disinfected out of breeding season but it is natural for the male odor to be more or less pronounced at mating time and that is the way in the wild state the herd gathers at eventime

NUBIAN DOE

when the kids are placed in a compact group, with the does laying down in a circle surrounding them, heads pointing outward, while the King of the herd serves as Sentinel giving a snort at the approach of an intruding animal such as mountain lions, wolves, coyotes, etc., who will find a foe of massed defenders at beginning, who usually come off victorious.

NUBIAN BUCK

# NUBIAN STANDARD

The Committee in charge recommends the following standard, which at present has not been officially adopted nor can it be until the next annual meeting of the Record Association though it does give a comprehensive idea as to points to be observed.

## *Descriptive Standard for Nubian Doe*

### HEAD 13 POINTS

Head of medium size, fine in contour, feminine; eyes large, clear, bright; broad between eyes; Cut 2 points for crooked face .......... Score    3 points

MUZZLE broad; strong nostrils, lips, jaws, "Under-shot" jaw (The lower jaw projecting beyond the upper jaw) permitted since considered no fault for Nubian head in England where this breed was "made" (P. 74. British Yearbook 1939) .... Score    2 points

Nose strongly convex ......... Score    3 points

Ears very long, also wide and pendulous, not too thick, hanging close to
  head ..................... Score    3 points

Poll—naturally hornless credit 2
  points; neatly disbudded without
  scurs, credit 1½ points; horns or
  large prominent sawed-off stubs, discredit 2 points ............. Score    2 points

## NECK 3 POINTS

Neck long and slender; preferred without wattles with no dewlap;
  smoothly jointed to head and shoulders; wattles, if present, evenly
  hung. Cut ½ point for wattles

                         Score    3 points

## BODY 22 POINTS

Shoulders slightly lower than hips;
  light and smoothly blended into
  body; withers thin and sharp; wide
  between shoulders; chest deep and
  wide between and back of front

legs; forelegs straight and strong

Score   6 points

BODY long with well sprung ribs and ample heart-girth; deep and well developed abdomen firmly held up; back straight and strong with broad long rump, only slightly sloping; hip bones wide apart; the thurls preferably somewhat wider apart than hip bones; vertebrae open, with prominent spinal processes

Score  10 points

HIND-LEGS clean, strong, straight, proportionate as to size, without tendency to being cow-hocked: hind quarters deep, somewhat concave behind, allowing plenty of width for udder, upright pasterns, feet true, pointing forward ......... Score   6 points

## MAMMARY SYSTEM 37 POINTS

UDDER globular, capacious attached over a large area; free from fleshiness, flexible and of fine texture; attached well forward at front, with no cavity between the two halves; at rear, well rounded, attached high and firmly, with no signs of being pendulous; halves even, with no indentation between halves and so smoothly joined as to present a blended rounded contour at rear

Score 25 points

TEATS symmetrical, pointing downwards and slightly forward; moderately long and of a size that can be comfortably held in the hand for milking; teats should be squarely set, wide apart, distinct from udder, not blending into udder; good orifice for milking strong stream of milk .................. Score 8 points

MILK VEINS large, long, prominent, tortuous, with lateral branches . Score 4 points

## SIZE AND GENERAL APPEARANCE 25 POINTS

SIZE—mature does; height at withers 30 inches and up; weight 130 lbs. and up ..................... Score 10 points

GENERAL APPEARANC—rangy, active, vigorous; hair short, soft, fine, glossy; skin loose and pliable. Any color, solid or parti-colored permitted. Toggenburg markings not desirable, discredit ½ point for such markings ............... Score 15 points

## Descriptive Standard for Nubian Buck

### HEAD 17 POINTS

Head, masculine and virile in proportion to the goat, and not excessively large: nose rather short and convex; eyes large, clear and bright; broad between eyes; ears long, wide and pendulous, not too thick ... Score   6 points

Muzzle broad; nostrils large and well developed; strong jaws. Lower jaw may or may not be "under-shot" (The lower jaw projecting beyond the upper jaw) .......... Score   6 points

Poll—naturally hornless credit 5 points; neatly disbudded without scurs, credit 3 points; horns or large prominent sawed-off stubs, discredit 5 points ................ Score   5 points

## NECK 5 POINTS

NECK strong and rather thick, with or without dewlap; preferred without wattles; wattles, if present, should be evenly hung. Cutting ½ point for wattles .............. Score  5 points

## BODY 38 POINTS

SHOULDERS usually slightly higher than hips, but no discredit if even with hips; strong and powerful, blending smoothly into body; withers thin and sharp; wide between shoulders; chest deep and wide between and back of forelegs; forelegs straight and strong with good bone development .................. Score 10 points

BODY long with well sprung ribs and ample heart-girth; deep and well developed abdomen firmly held up; back and rump strong, with rump only slightly sloping; hips wide apart .................... Score 18 points

HIND-LEGS straight, strong, facing for-

ward; wide between hocks, upright
pasterns, feet true ......... Score **10** points

## ORGANS OF REPRODUCTION 10 POINTS

ORGANS OF REPRODUCTION—well developed, showing breeding ability. Correct number of rudimentary teats, namely two .......... Score **10** points

## SIZE AND GENERAL APPEARANCE 30 POINTS

SIZE—mature bucks; height at withers 35 inches and up; weight 175 lbs. and up. .................. Score **10** points

GENERAL APPEARANCE—rangy, active, vigorous; hair short, soft, fine, glossy; skin loose and pliable. Any color, solid or parti-colored permitted. Toggenburg markings not desirable, discredit ½ point for such markings .............. ..Score **20** points

NOTE.—Buck kids may be judged by the same standard as mature bucks.

## Standard for Nubian Doe Kids, and Yearling Does Not in Milk

KIDS and YEARLINGS should be judged on the basis of the promise which they show of developing into mature does of high quality as described in MATURE DOES' STANDARD. This does not mean, however, that a kid with a precocious udder should be placed ahead of a kid with normal udder development. If a kid has the development normal to her age, she should receive full credit in her score on udder development. Similarly when milking yearlings are shown in the same class with yearling not in milk, no advantage should be given the milking yearling for her superior udder development as compared with the yearling not in milk.

## Disqualifications

Unsoundness of any kind, disease or deformity disqualify a goat in the show-ring. Any artificial means of removing or remedying any defects by cutting or filling under the skin disqualifies, with the exception of dehorning, disbudding, clipping and trimming the hair. Hermaphrodites and castrated goats are ineligible to show. Dying the coat is considered as fraud and deception and disqualifies the animal. A goat that does not show the characteristics of her breed is disqualified to place in the breed classes.

The preceding constitute the Majority Report for the Nubian Descriptive Standard.

The Minority Report concurs with the Majority except in the following particulars:

Under MUZZLE, in both the buck and doe Standard: "Undershot" jaw not permitted.

Under EARS; both buck and doe Standard: Large, wide and pendulous, not too thick, should have muscular power to lift out partially from side of face, showing alertness.

Under Kids and Yearlings, add: Milking yearlings should be given preference due to their visible proof of their ability. Preferably should be shown in separate class.

# ALPINE BREED

This breed is a built breed with more or less advantages as such. Many years back a number of French men of ambition, it is claimed, concluded they would establish a breed combining the attributes of production, size and stamina regardless of breed, color or horns and in consequence selected the best individuals obtainable. This selection coupled with development and culling rewarded them in the course of a few generations in a given type which has gained prominence in several countries and while importation of this breed to this Country was of more recent years than the Swiss varieties and the Nubian, it has demonstrated the wisdom of the founders in making strong and outstanding characteristics coupled with production. Like the Nubian this breed may be of any color or combination of colors and is making some advance in breeding away from horns. Another feature resembling the Nubian is the fact in each breed the number imported was limited necessitating interlocking blood lines though by avoiding direct inbreeding there seems

FRENCH ALPINE DOE

to be no noticable ill effect, such as increasing number of hermaphrodites which has always been set as one of the prime objections to inbreeding. Coupled with depletion of size, production and

stamina, the original selection and cultivation of
the latter item however, seems to have proven en-

FRENCH ALPINE BUCK

during and yet there may come a day when an infusion of new blood may be desirable even though
in keeping with American ambition we have made
the Alpine in this Country, under our system of

official test for advanced registry become the Alpine records of the World and not far in the rear of the leaders of some other varieties.

The high milk record of this breed is held by Little Hill Pierrette's Lady Penelope 47939. A.R. #217 which is 2929.0 lbs. milk during 10 months' full term official test and the high butter fat by LaSuise Tar Baby 48596 Adv. Reg. 433 which is 95.637 lbs. butter fat during an 8 months' 29 day official test, qualifying in less than full term.

Since the age of the does at time of test has much to do with the results it is rather difficult to make comparison but all are worthy of being regarded as Standard bearers for their respective breeds and should serve as a Standard to breeders to strive to equal and even surpass them and it is worthy of consideration that the various top figures given above are as of the time this book is written.

# ALPINE STANDARD

The Committee in charge recommends the following standard but to date this has not been officially adopted nor can it be until the next annual meeting of the Record Association though it does give a comprehensive idea as to points to be observed.

### *Descriptive Standard for French and Rock Alpine Doe*

#### HEAD 8 POINTS

Head of medium size, fine in contour, feminine; facial lines straight or dished; eyes large, clear, bright; broad between eyes; ears erect, medium length, standing forward at attention, cone and horned shaped
Score  3 points

MUZZLE broad; strong nostrils, lips, jaws .................... Score  3 points

POLL—naturally hornless credit 2 points; neatly disbudded without scurs, credit 1½ points; horns or large prominent sawed-off stubs, discredit 2 points ........ Score   2 points

## NECK 3 POINTS

NECK long, slender and graceful; preferably without wattles; without dewlap; strongly and smoothly joined to shoulders ....... Score   3 points

## BODY 22 POINTS

SHOULDERS level with, or slightly lower than hips; light and smoothly blended into body; withers thin and sharp; wide between shoulders; chest deep and wide between and back of front legs; forelegs straight and strong .............. Score   6 points

BODY long with well sprung ribs and ample heart-girth; deep and well developed abdomen firmly held up; back straight and strong and broad long rump, only slightly sloping;

hip bones wide apart, the thurls preferably somewhat wider apart than hip bones; vertebrae open, with prominent spinal processes

Score 10 points

HIND-LEGS clean, strong, straight, fine, proportionate as to size, without tendency to being cowhocked; hind quarters deep, somewhat concave behind, allowing plenty of width for udder, upright pasterns, feet true, pointing forward .... Score 6 points

## MAMMARY SYSTEM 37 POINTS

UDDER globular, capacious attached over a large area; free from fleshiness, flexible and of fine texture; attached well forward at front, with no cavity between the two halves; at rear, well rounded, attached high and firmly, with no signs of being pendulous; halves even, with no indentation between halves and so smoothly joined as to present a blended rounded contour at rear

Score 25 points

TEATS symmetrical, pointing downwards and slightly forward; moderately long and of a size that can be comfortably held in the hand for milking; teats should be squarely set, wide apart, distinct from udder, not blending into udder; good orifice for milking strong stream of milk .................. Score   8 points

MILK VEINS large, long, prominent, tortuous, with lateral branches

Score   4 points

## SIZE AND GENERAL APPEARANCE 30 POINTS

SIZE—mature does; height at withers 30 inches and up; weight 135 lbs. and up ..................... Score  10 points

GENERAL APPEARANCE—alert, stylish, graceful and deerlike; strong and vigorous; coat short or medium silky, lustrous; skin loose, pliable and supple ............. Score  15 points

COLOR—all colors permissible, including multiple robes. Color ranges from pure white through varying shades and tones of fawn, grey, piebald, and brown to black and showing various markings, shadings and combinations on the same animal. Other things being equal, prefer-ence should be given to animals showing the perfect type of color of best known Alpine varieties, as Cou Blanc, Cou Clair, Chamoisee and Sundgau. The Cou Blanc variety has white neck and shoulders, shad-ing to silver-grey, then to glossy black on the hind quarters, with grey or black markings about the head. The Cou Clair has grey, saf-fron or tan neck shading to gray, then glossy black on the hind quar-ters with grey or black markings about the head. The Chamoisee is tan, red, bay or brown with black markings on head or neck, black stripe down back and black legs. The Sundgau is glossy black with white markings on face and under-neath the body ........... Score  5 points

## *Descriptive for French and Rock Alpine Buck*

### HEAD 17 POINTS

Head, masculine and virile, clean in outline, strong but not coarse; facial lines straight or dished; eyes large, clear, bright; broad between eyes; ears small, cone shaped, erect, high on head, fine texture .. Score  6 points

MUZZLE broad; strong nostrils, lips, jaws ................... Score  6 points

POLL—naturally hornless credit 5 points; neatly disbudded without scurs, credit 1½ points; horns or large prominent sawed-off stubs, discredit 5 points .......... Score  5 points

### NECK 5 POINTS

NECK strong and heavy; preferably without dewlap; preferably without wattles ................ Score  5 points

### BODY 38 POINTS

SHOULDERS may be even with, or, at crest, slightly higher than hips; strong and powerful, blending smoothly into body; withers thin and sharp; wide between shoulders; chest deep and wide between and back of forelegs; forelegs straight and strong with good bone development .................... Score 10 points

BODY long with well sprung ribs and ample heart-girth; deep and well developed abdomen firmly held up; back and rump strong, with rump only slightly sloping; hips wide apart ................... Score 18 points

HIND-LEGS straight, strong, facing forward; wide between hocks; upright pasterns, feet true ........ Score 10 points

## ORGANS OF REPRODUCTION 10 POINTS

ORGANS OF REPRODUCTION—well developed, showing breeding ability. Correct number of rudimentary teats, namely two ............. Score 10 points

## SIZE AND GENERAL APPEARANCE 30 POINTS

SIZE—mature bucks; height at withers 35 inches and up; weight 175 lbs. and up ................. Score 10 points

GENERAL APPEARANCE—active, vigorous, rugged but not coarse; coat smooth and glossy; short or medium; skin thin and loose ... Score 15 points

COLOR—all colors permissible, including multiple robes. Other things being equal, preference is given to animals showing the perfect type of color of best known Alpine varieties, as Cou Blanc, Cou Clair. Chamoisee and Sundgau. For detailed description, see COLOR as per doe standard ......... Score 5 points

NOTE.—Buck kids may be judged by the same standard as mature bucks.

## Standard for French and Rock Alpine Doe Kids, and Yearling Does Not in Milk

KIDS and YEARLINGS should be judged on the basis of the promise which they show of developing into mature does of high quality as described in MATURE DOES' STANDARD. This does not mean, however, that a kid with a precocious udder should be placed ahead of a kid with normal udder development. If a kid has the development normal to her age, she should receive full credit in her score on udder development. Similarly when milking yearlings are shown in the same class with yearlings not in milk, no advantage should be given the milking yearling for her superior udder development as compared with the yearling not in milk.

## Disqualifications

Unsoundness of any kind, disease or deformity disqualify a goat in the show-ring. Any artificial means of removing or remedying any defects by cutting or filling under the skin disqualifies, with the exception of dehorning, disbudding, clipping and trimming the hair. Hermaphrodites and castrated goats are ineligible to show.

Dying the coat is considered as fraud and deception and disqualifies the animal. A goat that does not show the characteristics of her breed is disqualified to place in the breed classes.

The preceding pages constitute the Majority Report on the Description Standard for French and Rock Alpine breed.

The Minority Report concurred with the Majority except for the following:

Under Kids and Yearlings: Milking yearlings should be given preference due to the visible proof of their ability. Preferably should be shown in a separate class.

# MURCIENE BREED

This beautiful neat animal was introduced several years back from Spain. The predominating color is seal brown, short silky coat, largely hornless, with well developed udder and the milk usually is high in butter fat. Unfortunately, the original trio one buck and two does were not advantageously promoted either in the show ring nor were they properly advertised, hence many prominent breeders of other lines have never seen one. The registration of this breed halted for a time but later recovered.

Regretfully the breeders of this breed failed to reply to our request for pictures of their stock.

## NARSKA BREED

This is a large animal brought over from Norway long years ago but like the Murciene, not promoted though worthy of being given proper recognition. Their first registration included the original importation and their descendants. The predominating color is white with some variation.

The same failure to respond to our repeated requests for pictures of this breed forces us to omit illustration here.

# ROCK ALPINE BREED

This is an American built breed based upon the first two Swiss importations, then topped by continuous use of pure French Alpine bucks and bred up to a point recognized through resolution by the American Milk Goat Record Association as being worthy of being given a standing of independent breed reproducing true to type. Like the Nubian and French Alpine they are not confined to any one color but their type, size and general quality commands respect of any one well versed in good points in a high class milk goat.

In three successive issues of the Goat World we asked owners of this breed to send prints or cuts which like others were probably overlooked until too late for this book.

# SWISS ALPINE BREED

This breed is one of the latest importations to be established in this Country and its name designates its place of origin. The usual size is slightly less than some lines. The color is principally a rich seal brown. The coat short, sleek and glossy.

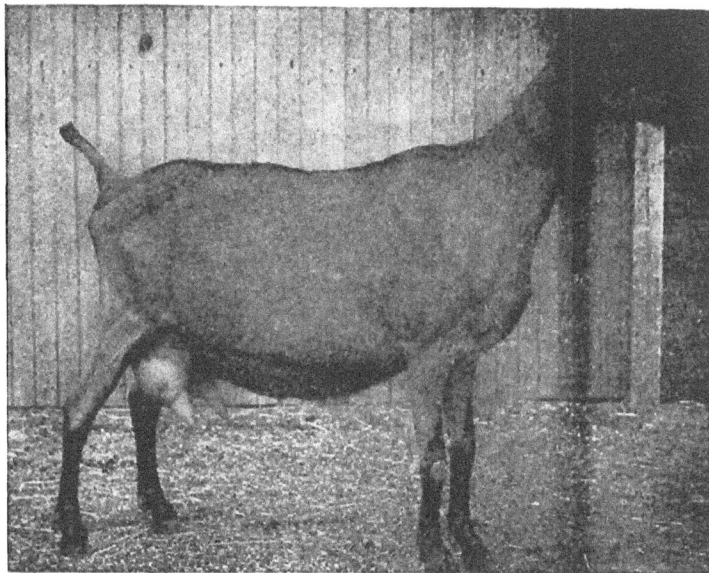

SWISS ALPINE DOE

## SWISS ALPINE BREED

Usually hornless and the milk, like that of all well kept goats, delicious, ample in quantity and quality for any reasonable purpose. Practically all breeders of any one breed think and assert theirs is the best, but the broadminded breeder, whether he breeds one or all varieties, grants the other varieties each have some outstanding special points. My own opinion is they are all good and any one good enough to meet my individual needs and desires and receive my unqualified endorsement.

SWISS ALPINE BUCK

## *Your Guide to Breeding.*

WHILE I hope that you will start your goat-keeping with an in-kid nanny, so that questions concerning breeding will not trouble you during your first few months in your new venture, it is best, I suppose, to take first things first, and deal in this chapter with the mating of goats.

As far as milk production only is concerned, mating a nanny with any handy billy will produce the desired result. But where it is also desired to rear kids and improve one's strain, rather more is involved. Only pedigree males from proved milking stock should be used.

Goat-keepers have been particularly fortunate in the matter of breeding in that, until the war, the Government subsidised a Stud Goat Scheme by which cottagers, small-holders, and artisan goat-owners could obtain the services of first-class pedigree stud males at fees not in excess of 4/——fees which in the ordinary way might be two or three guineas.

While it is unfortunate that, owing to the war, the Scheme has had to be suspended, the majority of stud goat owners are still allowing services on pretty well the same terms; and such stud goats are to be found in practically every district. You can start with the veriest scrub nanny and by breeding consistently to pedigree males can have gallon milkers of your own within a very few seasons.

Normally goats breed only during the season extending from the beginning of September to the end of February. The period the kids are carried is approximately five months or 21 weeks. When you buy an in-kid goat, then, find out exactly when she was mated, and you can tell within a few days when the kids will be born. The following table will help you in this respect.

I said previously that goats *normally* breed from September to February inclusive. Goats *can* be mated before

85

## BREEDING TIME-TABLE.

Breeding Season taken as from September 1st to February 28th. Period of gestation as 21 weeks. Allowance for leap year is made. Duration of Oestrum—2-3 days. Abortion most liable to occur at 5th, 9th, or 13th week.

| Service Date | September | 1 | 2 | 3 | 4 | 5 | 6 | 7 | 8 | 9 | 10 | 11 | 12 | 13 | 14 | 15 | 16 | 17 | 18 | 19 | 20 | 21 | 22 | 23 | 24 | 25 | 26 | 27 | 28 | 29 | 30 | |
| Kidding Date | January | 26 | 27 | 28 | 29 | 30 | 31 | Feb. 1 | 2 | 3 | 4 | 5 | 6 | 7 | 8 | 9 | 10 | 11 | 12 | 13 | 14 | 15 | 16 | 17 | 18 | 19 | 20 | 21 | 22 | 23 | 24 | |

| Service Date | October | 1 | 2 | 3 | 4 | 5 | 6 | 7 | 8 | 9 | 10 | 11 | 12 | 13 | 14 | 15 | 16 | 17 | 18 | 19 | 20 | 21 | 22 | 23 | 24 | 25 | 26 | 27 | 28 | 29 | 30 | 31 |
| Kidding Date | February | 25 | 26 | 27 | 28 | 29 | Mar. 1 | 2 | 3 | 4 | 5 | 6 | 7 | 8 | 9 | 10 | 11 | 12 | 13 | 14 | 15 | 16 | 17 | 18 | 19 | 20 | 21 | 22 | 23 | 24 | 25 | 26 |

| Service Date | November | 1 | 2 | 3 | 4 | 5 | 6 | 7 | 8 | 9 | 10 | 11 | 12 | 13 | 14 | 15 | 16 | 17 | 18 | 19 | 20 | 21 | 22 | 23 | 24 | 25 | 26 | 27 | 28 | 29 | 30 | |
| Kidding Date | March | 27 | 28 | 29 | 30 | 31 | April 1 | 2 | 3 | 4 | 5 | 6 | 7 | 8 | 9 | 10 | 11 | 12 | 13 | 14 | 15 | 16 | 17 | 18 | 19 | 20 | 21 | 22 | 23 | 24 | 25 | |

| Service Date | December | 1 | 2 | 3 | 4 | 5 | 6 | 7 | 8 | 9 | 10 | 11 | 12 | 13 | 14 | 15 | 16 | 17 | 18 | 19 | 20 | 21 | 22 | 23 | 24 | 25 | 26 | 27 | 28 | 29 | 30 | 31 |
| Kidding Date | April | 26 | 27 | 28 | 29 | 30 | May 1 | 2 | 3 | 4 | 5 | 6 | 7 | 8 | 9 | 10 | 11 | 12 | 13 | 14 | 15 | 16 | 17 | 18 | 19 | 20 | 21 | 22 | 23 | 24 | 25 | 26 |

| Service Date | January | 1 | 2 | 3 | 4 | 5 | 6 | 7 | 8 | 9 | 10 | 11 | 12 | 13 | 14 | 15 | 16 | 17 | 18 | 19 | 20 | 21 | 22 | 23 | 24 | 25 | 26 | 27 | 28 | 29 | 30 | 31 |
| Kidding Date | May | 27 | 28 | 29 | 30 | 31 | June 1 | 2 | 3 | 4 | 5 | 6 | 7 | 8 | 9 | 10 | 11 | 12 | 13 | 14 | 15 | 16 | 17 | 18 | 19 | 20 | 21 | 22 | 23 | 24 | 25 | 26 |

| Service Date | February | 1 | 2 | 3 | 4 | 5 | 6 | 7 | 8 | 9 | 10 | 11 | 12 | 13 | 14 | 15 | 16 | 17 | 18 | 19 | 20 | 21 | 22 | 23 | 24 | 25 | 26 | 27 | 28 | 29 | |
| Kidding Date | June | 27 | 28 | 29 | 30 | July 1 | 2 | 3 | 4 | 5 | 6 | 7 | 8 | 9 | 10 | 11 | 12 | 13 | 14 | 15 | 16 | 17 | 18 | 19 | 20 | 21 | 22 | 23 | 24 | 25 | |

and after this period, but " in season " signs are not so prominent, and the results are not so certain. Also, no two goats are alike. Some, coming early " in season," kid down in early spring (all to the good, the kids having the best " growing " weather before them); but others do not appear in any hurry, and in seeking their mates late in the season kid in late summer or early autumn—these are termed " winter milkers."

How to tell when a goat is " in season "? Here, again, no two goats are alike. Some, very quiet, and possibly shy ones, will show they are in season by a rapid shaking of their tail, a restlessness of manner, and sometimes a refusal of food. Those that are giving milk will drop their yield suddenly—a goat giving usually two or three pints will one morning give about half a pint! Of course, the yield returns after the " in season " period which may last from two to three days.

The noisy goats will bleat continuously, also shaking their tails, although it may be a full 24 hours before they are properly " in."

I think the best way to be sure is to walk quietly round and try to get a rear view of the " suspected one." If under the tail there is a swelling, red and sore looking, or bluish in colour, and a whitish discharge, or even transparent, the goat should be taken at once to the male, if you wish her mated.

Just a word of warning here to those who want to breed *good* goats. If in your locality, even as far as two miles distant, there should be an undesirable little scrub, do not leave your " in season " nanny loose or tethered. Either she will visit the undesirable one, or he, with that most wonderful, unerring animal instinct, will visit her. So watch carefully every morning until the goat is mated that no such accidents happen.

And when the goat has been to her chosen mate, still watch. She may " turn "—that is, come " in season " again a week after mating, three weeks (the usual time elapsing between each " season " until mated), at four, or even six weeks. Therefore a goat cannot be guaranteed " in-kid " until she has been six weeks mated without any further sign of " season."

It sounds a lot of bother, but it is not. Any good stock-keeper will have a " watchful eye " always on the stock—and goats are valuable stock.

# MATING SEASON

This is one of the most important points for careful consideration for if the start is wrong, all

REAL PRODUCERS

is wrong. However, even with a proper start so far as is possible no definite assurance can be safely assumed, for in the most carefully established herds one may observe occasionally what is termed a sport, meaning a throw back, to some ancestor even though remote. But if a careful research is instituted in the forebears on both the sire and dam sides, it may show ever so slight a tendency toward some weakness or defect in anatomy, or even in temperament, as well as pertains to production and reproduction, did exist and when united in the offspring was materially intensified. In such cases it is well to use even more than the customary precaution in mating such an animal in order to overcome such defects.

Illustrating this point, I recall a prominent breeder of heavy producers which brought him a comfortable living by sale of the milk in a city where he had established a successful business. He brought some of his largest and in his estimation most valuable animals to the State Fair feeling confident of success. The bones, being too light to support the body, bent resulting in not only unsightly legs but this defect continued through generations. When the awards were made and he was not successful he almost wept with chagrin and demanded an explanation. To that not only he but every exhibitor and onlooker present was entitled to know and no one accepting appointment to judge should fail to state clearly the points, credit and discredit entering into his decision, for if a show ring is not an educational

AT THE FAIR

branch of the Industry, it fails to function in its proper rank and the time to point out the features entering into the decision is before the class is dismissed from the ring, and it is a mistake to hold a post mortem later for it is almost a certainty there will be more misquotations than true quotations to the detriment of all concerned. Finally acting upon my suggestion the above case was greatly improved by discarding the offender. While this is one, it is by no means the only point to be attained in careful selection, for mating though it is one of the basic points for sound frame and body is definitely necessary for continued success and should not be regarded lightly.

A HOUSEHOLD ASSET

# BREEDING OF
# THE ANGORA GOAT.

ONE can learn very little about breeding the Angora goat from the Turk. As we know from Tchikacheff's work, which was published over fifty years ago, cold winters often killed many of the Angoras in Asia Minor, and the Turk then imported from more favored districts common bucks or does to breed to the Angora. This was before the great demand for mohair, occasioned by the increase in manufacturing plants at Bradford, England, caused the Turkish mohair raisers to resort to all manner of means to increase the supply of raw material.

To-day the Turk is treading in the paths of his forefathers. What was good enough for them, certainly ought to be good enough for him, so he reasons. He eats with his fingers, cooks on a brazier, sits on the floor, eats, drinks, sleeps and works all in the same room, and keeps his wives in seclusion.

When he comes to breeding the Angora he leaves that to his servants, if he be wealthy enough to have any. Most of the breeders cannot read or write. They have never traveled. They have no ambition, and they know nothing of the principles of selective breeding.

As a natural consequence the Angora goat of to-day has not improved, nor is he likely to improve under Turkish management. One large breeder who supplied bucks to some tributary country, said that he thought that it was a shame to castrate a buck, no matter how bad he might be. The Turk separates the bucks from the does at breeding season, as Asia Minor has cold weather late in the spring, and the danger of losing kids, if they come too early, is great. When the bucks are turned with the flock they are allowed to run until the next breeding season, and all of the bucks, regardless of quality or quantity, are allowed to run with the does.

When the first few Angoras arrived in America the natural procedure was to cross them upon the common short-haired goat of this country. It was a new industry, and many wanted to try the Angora. Very slowly the Angora, or the cross-bred animals were scattered over the United States. Stories were told of the wonderful things for which the mohair was used, and some supposedly reliable authorities quoted mohair at $8.00 a pound, as has been stated. Companies were started, and of course the supply of good Angoras, that is, goats which would shear about four pounds of mohair (worth at that time about seventy-five cents or a dollar a pound), was limited. Men bought any goat which had a trace of Angora blood in him as a thoroughbred Angora. A few years, however, demonstrated the fact that a common goat, with a little admixture of Angora blood, did not produce either the quality or the quantity of fleece wanted. Only a few of the more persistent breeders continued the experiment and their investigations. They sent and went to the home of the Angora, and brought more of the original animals to America. It took the American breeders about thirty years to find out just what the Angora goat was and how he should be handled. During that thirty years large flocks of common goats, which had been crossed with the Angora, and which might be properly termed "grade flocks," had been formed. Only a few thoroughbred flocks, that is, flocks of the original Angora, as he came from Turkey, were in existence.

## CROSSING WITH THE
## COMMON SHORT HAIRED GOAT.

By experience we have learned that the common short coarse haired goat can be crossed with the Angora goat, and that after sufficient crosses have been made, the cross-bred Angora so nearly resembles the thoroughbred that for all practical purposes he is an Angora. We have also learned that certain kinds of common goats respond rapidly to the infusion of Angora blood, and that others retain certain peculiarities of the common goat for generations. The Angora will not cross with sheep. For instance, a common goat with a long mane on the back, or tuft of

PASHA V—A True Breeder.

long hair behind the foreleg, or on the flank or the hip, will continue to perpetuate this long coarse hair on the offspring for generations, even though the best of Angora blood be infused. The color of the common goat is of some importance. A brown or reddish brown goat retains the reddish cast at the base of the mohair much longer than one of a bluish or bluish black color. It is equally true that a pure white mother may drop a colored kid occasionally. In Constantinople the mohair is graded into parcels containing red kemp, black kemp, etc. There it is the kemp which retains the color. As has been stated, there is also a breed of brown Angora goats, or at least mohair-producing goats, in Koniah in Asia Minor. Presuming, then, that one has a suitable common doe and a good Angora buck as a basis, the following may be deduced as relative changes in the different crosses:

The first cross, or half-blood Angora, will have a covering of short coarse common hair and a thin covering of mohair, which does not grow very long. If the animal were to be shorn, possibly a half pound of hair of a very inferior grade might be yielded. If this hair were to be offered to a manufacturer, he would class it as noil, and refer it to a carpet manufacturer, who would possibly pay ten or twelve cents a pound for it. The skin of the animal will be a little fluffy, and not suitable for fine goat skin trade.

It will not take a good polish after tanning, and it is not desirable for shoe leather. It will be worth about half as much as common goat skin. The meat of the animal will be a little better than that of the common goat, but it will be inferior to Angora venison. The animal will still be as prolific as the common goat. Twins and triplets will be a common occurrence. The kids will also be hardy. If one were to stop at this stage in breeding, he would have decreased the value of the skin of his goat without increasing the value of the animal.

The second cross, or the three-quarter blood Angora, will have a covering of short coarse common hair, especially noticeable on the back, belly, neck and hips. The mohair will now be fairly thickly set upon the sides of the animal, and of medium length, about seven inches long for a year's growth. If the animal were to be examined by a novice, he would be called an Angora from his general appearance. If shorn, he will yield about one, or one and a half pounds of hair, and the mohair manufacturer will pay about twelve or fifteen cents a pound for the material. The skin is valueless for rug, robe or trimming purposes, because of the coarse back and the scanty covering of mohair. It is fit for glove leather after tanning, but its value for this purpose is less than that of the common goat. The meat is more like Angora venison, and can be sold on the market as mutton. The animal is still prolific. From the second cross on, the grade goat rapidly assumes the characteristic of the Angora goat, but if for any reason poor bucks are used (an occasional animal without apparent reason retrogrades), the animal as rapidly resumes the characteristic of the common goat. Quite a percentage of colored kids will be dropped by does which are themselves white.

The third cross, or seven-eighths blood Angora, will still have the coarse back, a partially bare belly, coarse hips, and the neck will be insufficiently covered. The sides will be covered with good quality, long staple mohair, comparatively free from the coarse, dead underhair, or kemp. The animal will shear about two or three pounds of fair mohair, which will be worth from twenty to thirty cents a pound. This mohair will be fit to run through the combs, and the "top," or long mohair, free from kemp, will be used in the manufacture of plushes, braids, etc. The skin will have some value for rug, robe and trimming purposes. The meat will be juicy, palatable and salable as mutton.

The fourth cross, or fifteen-sixteenths blood Angora, will be hardly distinguishable from the average thoroughbred Angora. The coarse back will persist to some extent, and the hip will be plentifully covered with kemp. A good many of this grade will be poorly covered on the belly, and an occasional bare necked or off colored animal will be dropped. The animal will shear from two and a half to five pounds of mohair of good quality, which will be worth from twenty-five to thirty-five cents a pound. It will be from eight to twelve inches long at a year's growth, and it will be combed at the mill. It is fit for manufacturing into any of the goods for which mohair is used. The meat of the animal is rich, juicy, and free from the disagreeable qualities so often noticeable in mutton. If the animal be fed upon browse, the meat will have the flavor of venison. The tendency of the mothers to drop twins will be lessened, and it will be rather the exception for twins to be born. The kids will be rather delicate when dropped.

Subsequent crosses will tend to reduce the amount of kemp upon the animal and to improve the back. The question will now resolve itself into one of breeding for points. Bucks must be selected which cover the points the does need most, and by careful selection the grade flock will soon be indistinguishable from the thoroughbreds.

## METHODS USED IN AMERICA TO-DAY.

By gradual steps the original Angoras imported into America have been so improved, and the cross-bloods have been so highly graded that some of the American flocks equal the best Turkish flocks. America has many high-grade flocks, which, if it were not for the remaining coarse hair of the common goat, would be upon a par with the Turkish flocks. There are enough good goats in the country for a foundation stock, and a few years more of the careful, painstaking, selective breeding which is in progress throughout the United States to-day, will bring forth an Angora superior to the Turkish stock. Sections of the country modify the characteristics of the Angora. Probably climatic conditions, varieties of food and water, and certainly mental vigor of the owners is largely responsible for this. One man selects large, well formed, rapidly maturing goats and breeds for this type. It is surprising how soon his flocks assume this type. Another breeder works for fineness of fleece, regardless of size or shape of the animal, and he gets his points.

There has been much vagueness as to what points the breeder should try to produce. Some have claimed that the most profitable animal to raise was one producing heavy ringletty fleece, regardless of the quality of the fleece, except of course that it should be as free from kemp as possible. This day has passed. We know what the mohair is used for, and know how it is prepared for manufacturing. The future may change these uses or methods, but we know what we want now, and we know how to breed our goats to produce the most money per head for the present at least. Fashions vary, and the fashions vary the demand for certain grades of mohair. Coarse fibered, long staple, fine luster mohair possessing a great amount of tensile strength and elasticity will make good braid yarns, but if braid yarns are not in demand, such fiber is not the best for plush or dress yarns. Fine fibered, long staple, pliable, lustrous, easily spun yarn can be used for braid stuff, or at least part of the fleece will be heavy enough for this purpose, and the finer parts have such a variety of uses that they spin yarns which are always in demand. Looking at the question from the manufacturing standpoint, we see that the most staple product is the fine-fibered mohair. But a producer might have animals which would shear two and a half pounds average (the average of the Turkish flocks) of very fine mohair, while another grower might have animals which would shear four or five pounds average of coarse mohair. And even though the value per pound of the coarse mohair may be considerably less than that of the fine mohair, the grower owning the coarse haired heavy shearing Angoras will realize more money per head for his clip. The value also of the carcass and skin of the Angora is of importance. A heavy carcass and a large skin are of more value than a light carcass and a small skin.

If the Angora breeder would produce the animal which will yield the most money per head, he should aim to produce an animal which will shear the heaviest fleece of the most marketable mohair, regardless of fashions, and one which, when put upon the market, will dress the most possible pounds of desirable meat, and yield a readily marketable skin. There are not many such animals on the market to-day, but the time when there will be plenty is coming. We have the fineness of fiber; we have the density of weight of fleece; we have the covering of the animal and the size and stamina of the individual, and we have breeders who are endeavoring to unite combinations to produce the Angora of the future. But while we are without the ideal, one should choose that point which is hardest to attain, most necessary for the best paying animal, and work especially for that. That point is fineness of fiber, always remembering freeness from kemp. There are many large goats, many heavy shearing goats, but there are very few fine fibered comparatively free from kemp goats. One should not make the mistake of neglecting size and weight of fleece. There are few animals which will respond more rapidly to careful crossing than the Angora goat. A buck will usually stamp his individuality upon every kid, hence the necessity of carefully selecting breeding stock.

## GESTATION.

The period of gestation varies slightly with the individual, but the average may be approximately stated as one hundred and forty-seven days, or about five months. Both the bucks and the does have a breeding season, but this season may be changed or varied by different elements. As a rule the bucks commence to rut about July or August here in America, and the does soon after the time the bucks commence. Some bucks which have been allowed to run with the does all of the time, never cease rutting, and the does conceive about every six months. The does come in heat about every fourteen days, and remain in this condition for about three days. If the bucks are allowed to run with the does, one buck should be used for about every fifty does. If the buck is only allowed to serve the doe once, a grown animal will serve one hundred and fifty does in forty days without permanent injury to himself. The does conceive at about the age of seven months, and the bucks breed at about the same age, but the wise breeder will not sacrifice the individual by interfering with its development. Both the buck and the doe should not be bred until they are at least a year old. The bucks should be fed at breeding season, and if one has a sufficient number of bucks, it is well to turn the bucks with the does in relays. It is advisable to have the kids start coming slowly, so that one may get new men trained to handle them properly. One or two bucks turned with a flock of a thousand does for a few days, and then removed and allowed to rest, and a new relay of three or more bucks turned with the does, to be removed in a few days, and a new re-

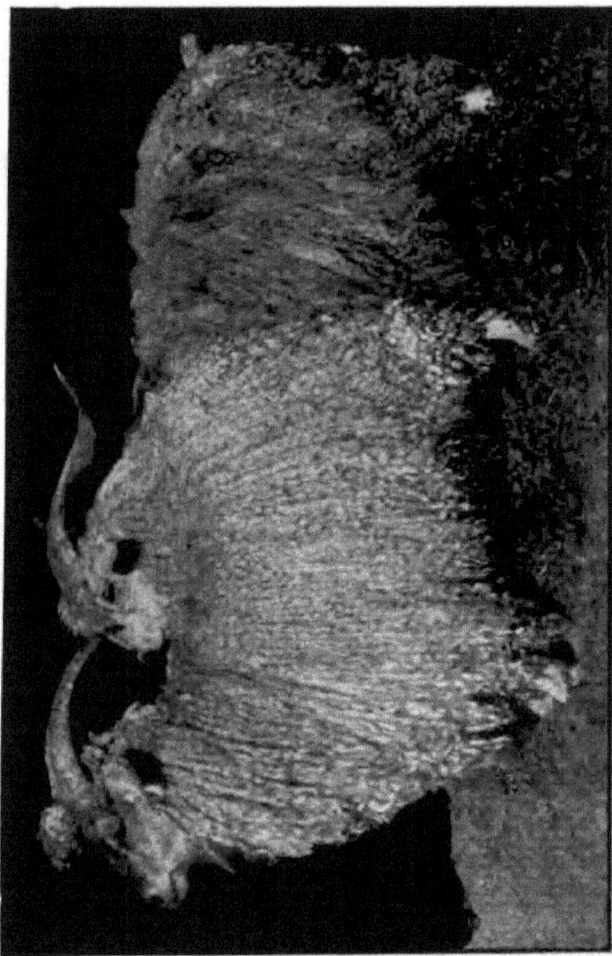

PASHA V AND BISMARCK.

American bred bucks, Bismarck shearing 12 pounds, was the sire of the grand champion buck at the St. Louis World's Fair, 1904.

lay being introduced into the flock, will do more satisfactory work than they would if all of the bucks were turned in at one time. The same principle can be applied to smaller flocks. The does should be protected from cold storms or rough handling when they are heavy with kid, else they are liable to abort. If for any unusual cause the doe aborts one season, there is no reason why she will not carry her kid until full term another time, and experience has proven that she will.

## BREEDING OF REGISTERED STOCK.

The breeding of registered stock, or stock of known ancestry, requires much care and quite different handling. Both the does and the bucks must be marked with an ear tag, brand, tattoo number, or some other permanent individual mark, and the kids should be marked at birth. Fifty known does may be put in a pasture or pen and a known buck put with them. He should be allowed to run with them at least forty days. After this the does may be collected into a flock and several bucks turned with them, but only the kids which are dropped from a known buck are fit for record.

A more accurate method, and one which can be used with a large flock, is to place the bucks in a corral adjoining the one used by the does at night. The does should be brought into their corral early in the evening, and all of those in heat will work aloug the fence next to the bucks. The doe in heat can be caught and the number taken and recorded in a book. She is then placed in a small pen with a buck and his number is recorded with hers, together with the date. If the doe does not conceive, she can be put with the same buck again at a later date, and one has approximate knowledge of when she should drop her kid. In this manner a buck will serve about two or three does in the evening, and one or two in the morning. The kid is marked at birth and the number recorded after that of the mother. The breeding of recorded stock is of value only for special reasons, and is not advisable with large flocks, as it is expensive.

# Breeding—Good and Bad

THE lactation period of most does lasts from eight to ten months after freshening, although it is not unusual for a doe to continue milking for a year or more. The breeding season, however, is limited to the fall and winter months. Does come in heat from September through March, with occasionally a doe ready for breeding in August, and on rare occasions even in July with Nubians. For this reason most goat owners breed yearly to be sure that a doe does not go dry during the months when normally they do not come in heat. If you have two does, their breeding can be so spaced as to keep you supplied with a continuous flow of milk over the year. The first breeding may be in August or September so that the doe will freshen five months later, in January or February; the second doe bred in November or December, to freshen in May or June. This will give you the heaviest supply of milk during the summer months when more milk is consumed, but will keep some milk available even during the cold months, which is an important matter if the milk is used for infant feeding.

The indications of a doe in heat are several, although all may not be evident in a particular doe and it is sometimes difficult for the novice to recognize them. Perhaps the most noticeable is a continuous twitching of the tail which becomes more pronounced if the hand is passed along the doe's back. With this may go a loss of appetite, bleating, redness and swelling around

This common goat (*upper opposite page*), of little value as a milker, was bred to a purebred Saanen Buck. Her daughter (*lower opposite page*), is a half blood. Again a pure-bred sire was used. Here you see the result—a ¾ blood. Notice the improvement in type.

the vulva and sometimes a mucous discharge. A doe in heat is apt to annoy her companion goats by rubbing against them. Watch for these indications with the coming of fall and if you desire to breed the doe hustle her off to a buck, which you have previously located, without delay. Although the heat period may last two or three days, it may also last but a few hours. If you are not prepared to breed her, she will again come in heat in twenty-one days (sometimes sooner) until bred. When you take her to the buck don't just leave her and go off on other business. Actual mating takes less than a minute, and it is better to wait and be sure that she is served and then take her away. Repeated service is of no benefit to the doe and a distinct disadvantage to the buck—a waste of energy.

When the doe has been served, she will indicate it by arching her body and perhaps giving off a mucouslike discharge.

Make a record of the date and watch for a return of the heat symptoms in twenty-one days, or even sooner. Recurrence of these would indicate that the doe is *not* pregnant and service should be repeated. If in twenty-one days she does not come in heat, it is safe to assume that she is bred. Within a month or two she will show definite signs of being with kid. Count off five months (146–153 days) from the date of service and make a record of this, too, as the date on which the kids should be born.

Sometimes, for some inexplicable reason, a doe does not seem to come in heat. This occurred with a doe brought to me for service in February. All season she had apparently not come in heat and her owner was concerned that breeding might be missed. The doe was taken to the buck each day for several weeks, but showed no interest in him. At the end of March the veterinarian inoculated her with stilbesterol. The following day she was very obviously in heat and was served by the buck. On the other hand, late in April a man hurried over with a doe to be bred. He was sure that she was in heat—her tail twitched constantly. Although doubtful, since the season was so late, I took her to the buck without result. Probably the tail twitching was due to the irritation of tiny gnats that come so often with the early spring days. However, we did discover that she had a plentiful supply of lice and suggested treatment so that the trip wasn't wholly a futile one.

Opinions differ as to the best age for breeding young stock, some breeders advocating early breeding if the doe is well developed and vigorous. Others prefer to wait until she is in her second year—about eighteen months old, so that she will be two years old at the time of first freshening.

Inbreeding—the mating of does and bucks of the same blood, such as father and daughter, mother and son, brother and sister —and linebreeding (the mating of closely related animals, but showing unrelated blood in either parent, such as grandfather and granddaughter sired by an unrelated buck, or grandmother

110

and grandson whose dam is of another strain) are branches of breeding that the amateur would best leave to the experienced breeder. It requires scientific knowledge of the laws of breeding and heredity, keen observation and understanding of the good and bad points of the animals involved, time and money to pursue them. Such breeding naturally results in concentration of certain characteristics possessed by the strain, but one may only succeed in emphasizing unsuspected weaknesses that would have been offset with outbreeding.

It is desirable, however, that your doe be mated to a buck of her same breed, and you should select the buck with the same care as you would use in purchasing an animal. See what his background is, what kind of daughters he has produced, and if he is healthy and vigorous himself. Too many people undo the work of years of conscientious effort on the part of good breeders by taking their does to the nearest buck with the explanation that they just want to freshen them. Later, the appeal of the doe kids is so strong that they either retain them or give them to some friend who might better have had a kid with the true characteristics of the mother's breed, not just a little mongrel, however lovable. With a registered buck of the mother's breed as her sire, the little doe should be a step ahead of her mother in appearance and production.

After breeding, the milking doe will gradually give less and less milk until after three months you will probably find it unnecessary to milk her. If, however, she continues to produce milk she should be dried off. This may be done gradually by omitting the evening milking, then milking every other day, then increasing the intervals between milking until the flow of milk has ceased. Or it may be done abruptly by just not milking the doe. In following the latter method it is necessary to give careful attention to her udder and to milk it out if necessary to give her relief in case it becomes crowded with milk. Occasionally, a doe who is a very heavy milker is difficult to dry off, but it must be done, for a twenty-four hour shift producing milk and nourishing her unborn kids is too severe a strain on the prospective mother and likely to end unhappily.

When the doe has been dried off, a fitting ration (obtainable from most grain companies) should be fed, or the usual grain ration should be lessened gradually and augmented with bran to supply bulk, so that by the time of kidding the doe is getting about half bran in her grain allowance. The hay feeding should be generous, but a leaner hay than alfalfa or clover used. She should have warm water two or three times daily. Her daily exercise should be continued, but care must be taken to protect her from incidents that might frighten her and cause running or bumping. A gentle word and her name spoken is often all that is required to calm a nervous doe.

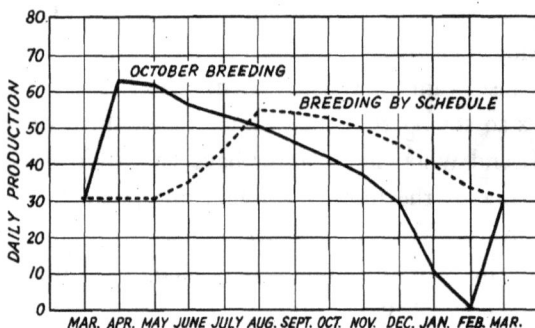

### Breed for Continuous Production

The goat industry's greatest practical difficulty is probably the seasonal breeding period. The normal breeding season is from Aug.–Jan. About ¾ of the goats in the U.S. are bred in the months of Sept., Oct., and Nov. and freshen in Feb., Mar., and April. Thus, production during late fall and early winter in most herds is very low. This is a serious handicap for any dairy in building a group of permanent customers.

Although the Dept. of Agr. at Beltsville, Md., has had some success in overcoming this problem by the use of hormones, no complete solution has been found. There does appear to be a greater than average range in the breeding season for Nubians.

In spite of this handicap much can be done through careful planning to smooth out the seasonal production curve of a herd. The graph illustrates these possibilities. The solid line represents production for a herd in which all goats are bred around Oct. 1, making them freshen in early Mar.

The following plan could be followed:

1. Breed 1/4 of the herd as soon as possible—by Sept. 1 to freshen before Feb. 1.

2. Breed 1/4 (the least persistent producers of those left) in Oct. to freshen in Mar.

3. Breed 1/4 (persistent producers) in Nov. to freshen in April.

4. Breed 1/4 (persistent producers) after Dec. 15 to freshen in late May or early June.

Groups 3 and 4 would give considerable milk during the usual months of shortage and would not dry off until after Group 1 had freshened. Such a breeding schedule would give a herd production curve similar to the broken line in the graph. A sizeable production would be secured even in the months of Dec., Jan., and Feb. This would be of great help in building a permanent trade.

# The Buck

WILL you need a buck?

If the herd is small, say less than ten or a dozen does, it is hardly advisable to have a buck, unless you live in an isolated section and can find no good buck of the same breed as your does available. With the exception that he doesn't need to be milked, the buck must receive the same care as the does. He must have his own quarters, weatherproof, well-lighted and comfortable, and far enough removed from the does so that he is not disturbed during the breeding season by their nearness. He should have his own exercise yard.

The buck needs brushing, hoof trimming, an occasional washing, regular feeding and watering just as the does do, and his pen must be cleaned regularly. People are oftentimes inclined to neglect the buck, giving him scant attention aside from his feeding, which has much to do with the reputation he gets of being an evil-smelling creature. There is no denying that even the best kept bucks have an odor, especially insistent during the mating season, but with reasonable care this may be kept to an inoffensive minimum—certainly much below the mark which announces his presence 100 yards away.

If you intend to go in for goat breeding you will, of course, need your own buck. Whether your does are grades or pure-breds, the buck should be a registered animal. Only with such a sire can you hope to retain the breed characteristics of your stock and improve it with each generation. You may prefer to purchase a mature buck who has given proof of his ability to produce daughters that are good milkers, or you may secure a buck kid and raise him yourself. For women goatkeepers the latter is the better plan for the young buck can be brought up in kindly fashion so that he will be easy to manage at all times.

In selecting a buck kid use the same precautions as in the selection of your does. Scrutinize his blood lines carefully and his immediate family—dam, sire and sisters if possible. Select a kid with good bone structure and a good, broad chest. Even a young buck will show quality if he has it. The price of a buck kid is, naturally, much less than that of a mature, proven buck. You do take some chance as to his potency, but a buck kid matures rapidly, and at six months is old enough for limited service. The chance is worth taking, for the molding of the buck's disposition is entirely in your hands. To my mind a buck who must be approached with pitchfork in hand or who wears a ring in his nose is proof of improper early handling, for goats are almost without exception fond of people and very responsive to kind treatment.

When the first milker I had, a Toggenburg doe, was ready for breeding I located a topnotch Toggenburg buck for service. A transcript of his pedigree was sent and I went to see him. I was taken into the barn and in the far end, amid a great rattling of chains, a very impressive looking animal was lowered on a platform. I stood back in fear—it was like a visit to the Spanish Inquisition. The buck was a fine animal and my doe was bred, but I stayed securely in the car until she was brought out to me.

When my herd was large enough to need a buck I secured a three-year-old. Later I learned that his owner had been afraid of him and he too had kept the buck chained at all times. As a result he didn't trust people and could never be fully trusted himself nor allowed complete freedom. He had a good-sized corral, but when he grazed in the open he had to be tied. Before venturing into his pen I always tied him at the far end of the corral. While he never injured anyone, he frightened a good many and he always required a watchful eye. Many times when

Toggenburg buck

I brushed him, although he liked to be brushed, he would send the brush hurtling through the air.

Since then I have raised several buck kids and not one has shown a disposition to be anything but gentle and friendly. Perhaps I have just been lucky, but I have known several women who have handled their own bucks and all have had the same good fortune. My present buck, now in his fourth year, is given complete freedom whenever possible. In good weather the door of his house is open and he can come and go in the field at any time when the does are not there. When they are in the field he is closed in his own yard. As soon as he hears me moving about in the morning, he comes down along the fence and calls. Whenever I want him I call his name and he answers or comes to see what is wanted. He receives the same care as the does—is fed and watered regularly, brushed, washed on occasion, his hoofs trimmed and his head scratched when he invites it. After

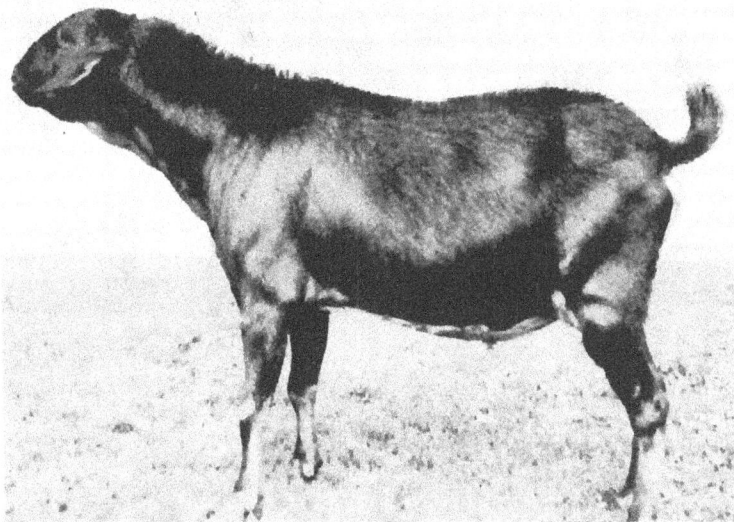

Nubian buck

a doe is served she is taken away immediately—a severe test of a buck's amiability—and he makes no protest. He does not live in lonely solitude, but has a wether (a castrated buck) of his own age for companion, brought up with him from baby days. Because he is not obliged to live in loneliness, he is quite devoid of the objectionable habits so often found in bucks. At times he and the wether have their differences, usually during the breeding season, but their house has two pens separated by a solid partition with a drop gate and, when it seems advisable, the gate is put in place and they are separated, yet side by side for companionship. When morning comes, they are once more good friends and glad to be together in the field.

If you decide on a buck kid be careful that he does not remain too long with the does or doe kids. As soon as he begins to get "bucky" give him his own quarters, for at three months of age a young buck is a potential father. But give him a playmate,

Saanen buck

or let him spend as much time as possible in your company. Do not, however, play with him or allow the children or the dog to push him about. It is amusing to see these kids rear up and butt in play, but it ruins them for later handling. You may train the buck to harness and use him for hauling manure, grass cutting and other work. It's good exercise for him—but he must respect you—and he won't if you become his playmate. Work quietly about him and don't startle him—always speak to him when you come into his house, and never beat him. A goat cannot be trained with the lash as can a dog into a cowering creature. He can just be made treacherous by such handling.

During the first season the young buck should be used only infrequently for service with long rest periods between matings. An older buck can serve thirty or more does during a season

A young buck or wether may be trained to harness.

and still keep in excellent condition if the breedings are distributed fairly evenly through the breeding months and limited to two a day. He should be well fed with plenty of good hay and a daily ration of grain, a pound or more, depending on your own observation of his body needs—enough to keep him in good flesh but not fat.

# FEED AND FEEDING

There is the common farm feed composed of grain and hay which is life giving and strength giving, but in order to obtain best results a truly balanced feed is obtainable in almost any locality through the instruction of your County Agricultural Agent, who knows or should know the percentages, proportions, etc. necessary. Then too, there are various commercial feeds made from formulas approved by government and state authorities and obtainable through your local feed stores, or if not it is a simple matter to order from the home office of the factory or its nearest distributing point, thus minimizing transportation cost. At all times it pays to use strictly first class feeds for a penny wise economy may prove a dollar extravagance.

As to hay, it is doubtful if ever the time will come when all feeders will agree upon any one variety, but safe it is to use alfalfa, lespedezia, alsike, red clover. Where any of these are run

way the stems as well as leaves will be eaten thus
through a cutting box in lengths of ¾ to 1 inch

FEED TIME EXPECTANCY

it will be found to be a great saving, for in that
saving what is often a wasteful procedure to over-
feed, thinking to supply amount sufficient for the
goat to pick out the more choice parts, and that

AN IDEAL INTERIOR PLAN

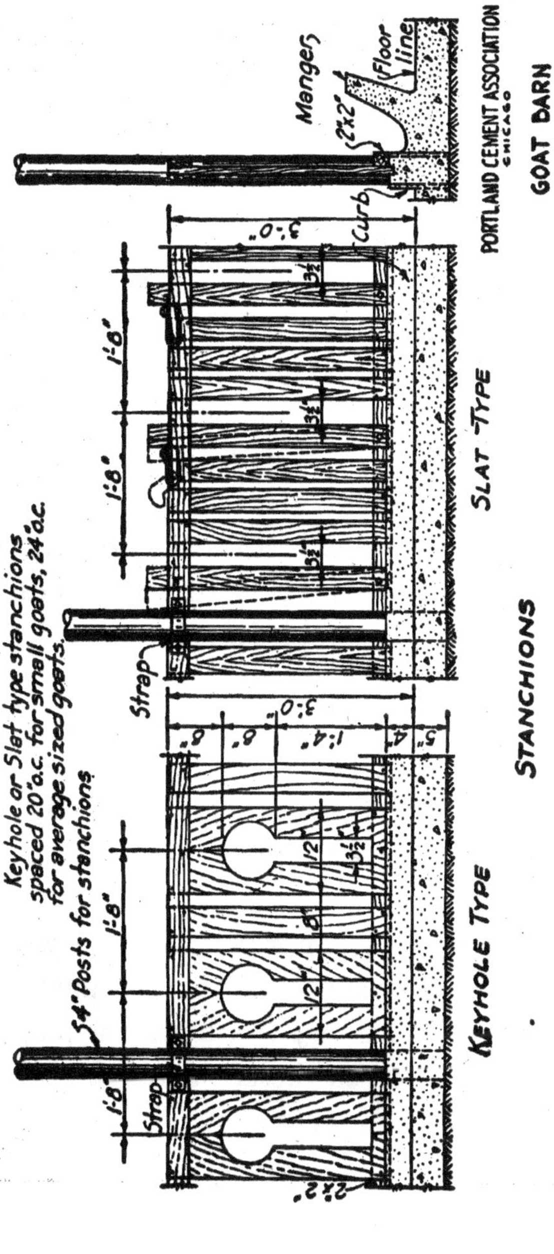

STANCHIONS

TAKE YOUR CHOICE

PORTLAND CEMENT ASSOCIATION
CHICAGO

GOAT BARN

SLAT TYPE

KEYHOLE TYPE

Mangers

Floor line

Curb

Strap

Keyhole or Slat type stanchions spaced 20" o.c. for small goats, 24" o.c. for average sized goats.

5½" Posts for stanchions

3'-0"

1'-8"

1'-6"

1'-4"

2"x2"

means when feed is pulled out of the manger and stepped upon, it will not be eaten for the fastidiousness of the doe in eating and drinking is well known.

While we have lightly touched upon this subject of feeding there is a vital point, the value of which can hardly be estimated, causing the oft repeated question the writer is called upon to answer viz: Why is it the same breed, the same feed, the same locality making it practically impossible to determine the cause of one herd proving so superior to another?

Well, you need only to call upon those two herds or if they belong to exhibitors you can even more quickly discuss the difference and definitely decide upon the cause for the more successful herd in barn or in the show ring will demonstrate the result of kind, gentle handling.

This does not necessarily call for the absurd extreme, which I have witnessed upon the part of an exhibitor who for effect will go through almost disgusting effort to impress the judge with silly and absurd words addressed to the animal even to the extreme of kissing, which with a judge may prove detrimental to the actor, though of course should not. The animal under full control,

# IMPROVED MILK GOATS

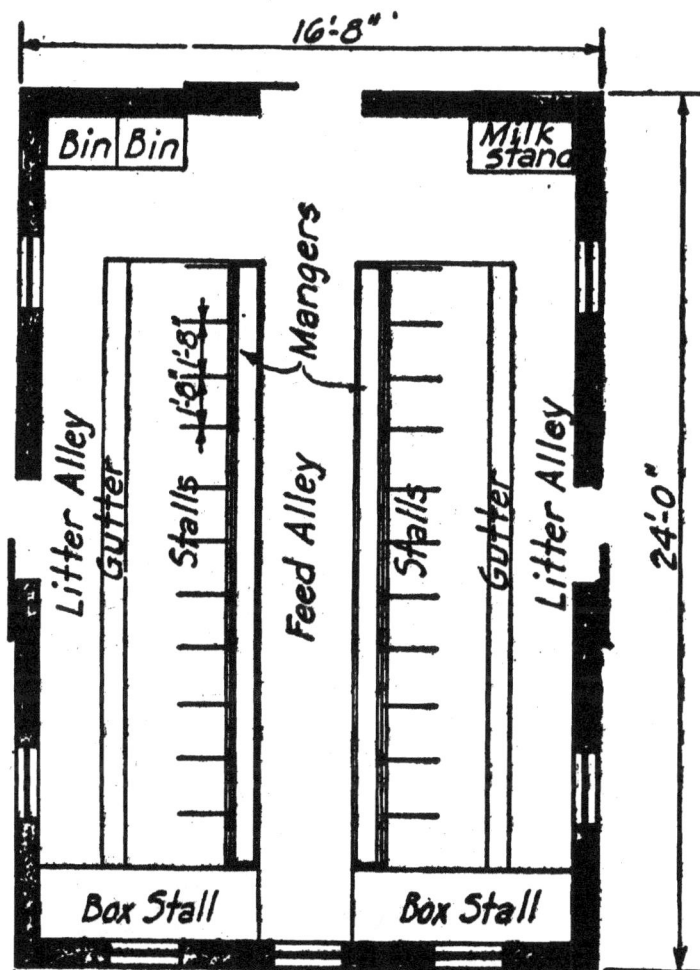

PLAN

INTERIOR PLAN SMALL STABLE

through gentle handling will not need to be fussed at, punished or scolded as some of the opposite extreme may resort to in the ring. But all in all,

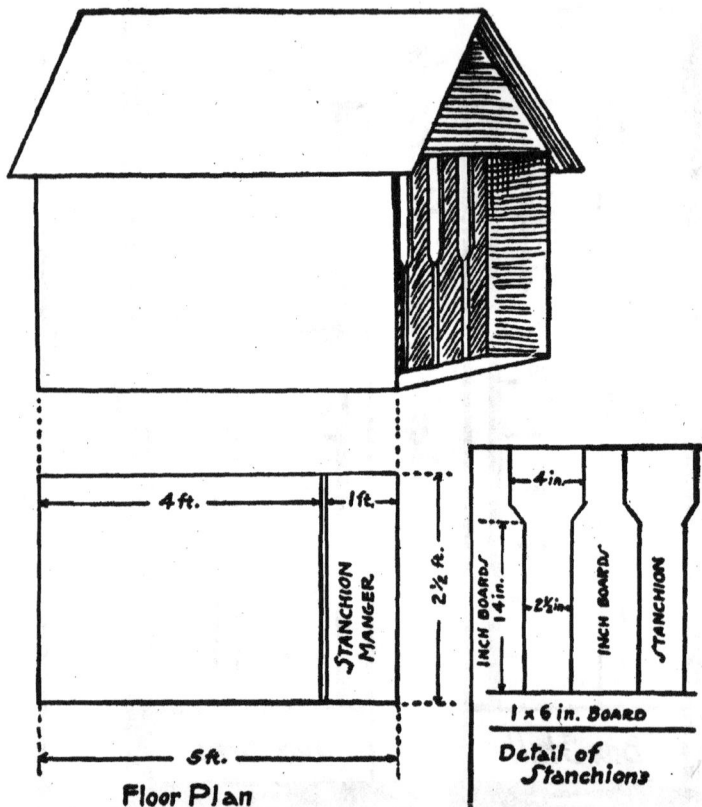

**Floor Plan**

4 ft. — 1 ft.

STANCHION MANGER

2½ ft.

5 ft.

INCH BOARDS 14 in.

4 in.

2½ in.

INCH BOARDS

STANCHION

1 x 6 in. BOARD

**Detail of Stanchions**

The bottoms of the boards forming the stanchions are fastened to the front of a 1 in. board, 6 in. wide, fastened crosswise of the box 1 ft. from the front end, forming a "toe-hold" for kids to climb on and manger for hay in front.

**EXTERIOR AND DIMENSIONS**

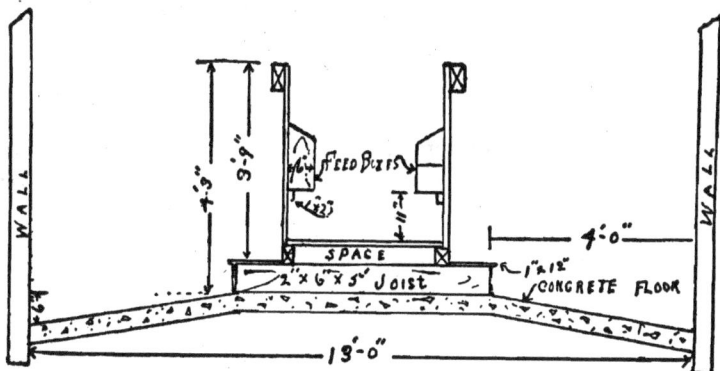

END VIEW OF GOAT STANCHIONS

**PROPER DRAINAGE FOR FLOOR**

the good feeder of good feed, discretely served according to actual requirements at regular intervals, will install a confidence in the goat which requires no fictitious actions to gain confidence of the judge or at home to the prospective purchaser.

In brief, use as much common sense as possible with which both you and I are endowed.

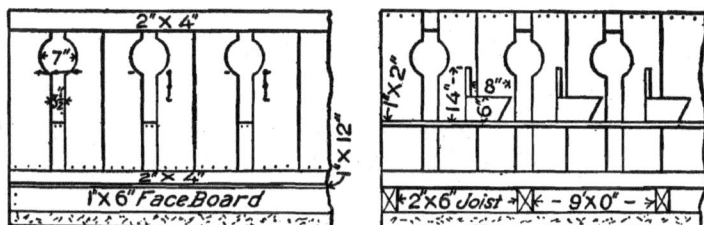

**FEEDING ARRANGEMENT**

# Feeding the Kids

NEWBORN KIDS may be left with the mother to get their milk directly from her udder, or they may be hand fed by bottle or pan. Most people prefer pan feeding. It takes less time, there is no breakage or tearing of rubber nipples, and the pans are more easily cleaned than the bottles.

Make your decision about the method you will use before the kids are born. It is extremely difficult to change a kid's feeding habit after she has had her first milk from the original source. Some people leave the kids with the mother for the first few feedings, to be sure that they get the colostrum, the first milk, and then shift to pan or bottle feeding, but my experience with this method has brought plenty of headaches.

I purchased a week-old buck, nursed by his mother, and spent hours trying to feed him from a pan with dismal, heartbreaking failure. Then I tried a baby's nursing bottle, holding it in his mouth and tilting his head so that gravity forced him to swallow. He was a beautiful, large kid, but grew thin and developed diarrhea, and was most forlorn. Finally I put a fresh doe on the bench and held her teat in his mouth, gently squeezing the milk into his mouth. He took it eagerly. By holding my hand on the doe's udder she thought that I was milking her and we all got on nicely, but I never could get that kid to feed from a pan or bottle.

If the kids are does and the goat milk is needed for human food it is best to hand feed them. Thus the supply of milk can

Where the milk supply is not needed, it's easiest to let the kids nurse. Breeders who don't sell milk usually follow this method.

be controlled and after three or four weeks on mother's milk they may be fed a substitute. Hand feeding has the added advantage that the kids are ready for sale and their feeding no problem to their new owner.

As soon as the newborn kid begins to nudge around its mother and would seem to be ready for food place your milk container in a pan of quite warm water and milk about a half-pint into it. Transfer this to a measuring cup, also warm, or if the doe is accustomed to being milked and will stand quiet, and your aim is good, set the measuring cup in the warm water and milk directly into it. Feed this to the kid at once, while quite warm. It should be 100° F. and you should use a thermometer for accuracy. You will find that the small circumference of the cup helps to keep the kid's head over the milk and she will

If goat milk is being used by the family—or sold—it's best to teach kids to drink from a pan. Skim milk or "calf" starter may be substituted for whole milk.

drink it down to the last drop. With a larger container she is apt to lose her place and grope about desperately, splashing and spilling the milk. In a week's time you may use a pan with better success.

A little kid is sometimes slow in taking her first feeding. If she isn't disposed to take the milk, cool it and place it in the refrigerator. In an hour or two reheat the milk to 100° F. and try again. Don't throw this first milk away for it contains properties very essential for cleansing the intestinal tract of the kid and antibodies to protect it from disease. It is a yellow milk sometimes quite thick and you may find when you are ready to use it again that you will need to thin it with boiled (not boiling) water. Be very careful in heating this colostrum milk and don't place it over direct heat. It is best to set your container in hot water and watch it closely. Too much heat will cause the milk to set like custard so that it cannot be used.

Unfortunately, little buck kids, unless they are of excellent breeding and are to be used for herd sires, seem to have no place in the world. They may be raised for driving goats or for pets, but oftentimes as they grow older their lot in life isn't a happy one. They are given to this one or that one, neglected and at times even abused. They are so intelligent and sweet that it is hard to think of butchering them, but it would seem that they are destined for food. The simplest and least troublesome way of feeding these kids is to leave them with the mother to help themselves. It is not safe, however, to assume that they are taking all the milk the mother is producing, especially during the first few days, and each day any excess should be milked out to avoid udder trouble and to keep up the doe's production. Also to prevent a lop-sided development of the udder, see that the kids take the milk from both sides.

By the time the buck kids are four to eight weeks old they are ready for butchering, and particularly if available at Easter-time there is a market waiting for them at a price ranging from $5 up, depending on their age and the demand. People who buy these buck kids for food prefer that they be milk fed exclusively —no hay or grain permitted—although there is no insistence that they be castrated. However, if after eight weeks they have not been sold it is wise to have them castrated, particularly if they are kept with the does or young doe kids, for they show an early interest in the does and even at three months are *capable of service.*

The most commonly used substitute for natural goat milk in kid feeding is powdered skim milk. The powder is added to cold water—¾ cup to a quart—and mixed with a rotary egg beater. This remade milk should be added to goat milk and introduced to the kid gradually, the amount of goat milk reduced and the skim milk increased each day until the goat milk has been wholly eliminated, if none can be spared for the kid's

Kids may also be fed by bottle. The milk should be warmed
to prevent scours.

feeding.

For the first few weeks the kid should be fed frequently, at
least four times daily, a half-pint at a feeding. Remember, the
milk should be warm. When the kid takes her feedings eagerly
they may be reduced to three daily and the amount increased.
At about a week old the kid will begin to nibble at hay and it
should be kept constantly before her, preferably at a little height
above her head so that she must reach for it. Also at about this

time try her with a fine grain mixture such as calf starter. As her appetite for hay and grain increases her milk feedings may be reduced to two a day and at three or four months may be discontinued entirely. A salt brick should be handy for her to lick. She should be given warm water, but not left to take her fill of it. Kids are greedy little creatures and will drink to the bursting point unless supervised.

Always have the utensils for the kid feeding clean, the milk warm and the bedding dry—which means frequent changing for kids do a great deal of wetting. Give them opportunity for exercise—a box or platform to jump on—fresh air and sunshine, and keep them warm, and they will be strong and healthy.

# Feeding for More Milk

BEFORE the goats arrive have their feed on hand. Otherwise you will find yourself in the predicament of Old Mother Hubbard with a bare larder and hungry animals, for goat food can't be had at the corner grocery store. Some people have the idea that a goat can be staked out to eat up the green stuff on the place and everything will be lovely with a full milk pail at the end of the day. This is a long way from the fact. Goats left to forage entirely for their food are brush goats and are not expected to produce milk.

To make milk, dairy goats need grain and hay winter, summer and all the year. When pasture is good they will eat very little hay, but before starting out in the morning should be given a light feeding to sustain them until the dew has dried from the growing things. This breakfast of hay will also minimize the danger of their eating poisonous plants in the pasture which might attract them if they are very hungry. On stormy days when they must remain indoors—for goats dislike rain and wet weather—they need hay, and during the night they will nibble on hay with relish and benefit.

The leguminous hays—alfalfa, clover—are the best milk producers and also the most expensive to buy. If you can afford to set your goats' table with alfalfa they will pay you dividends in increased milk production and sleek, shining coats and healthy bodies. However, a goat will eat and thrive on almost any sweet smelling, properly cured hay provided it is not stalky. If there

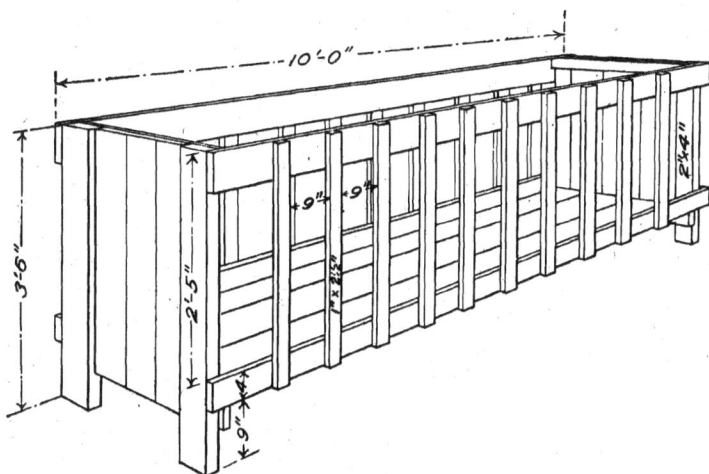

Combination hay and grain rack with bottom for feeding both roughage and grain.

are bits of brush and leaves mixed in she will relish it even more.

During the season when pasture is scant a goat needs at least three pounds of hay daily, but the safe procedure is to allow her all the hay she will eat—not in one daily feeding, but at intervals—three or four feedings during the day, not too much at a time. In this way she will waste less. A goat is much like a child and will pick out the choice bits first, scattering the less desirable pieces in her efforts to find what she likes best. She has a reputation of being very wasteful of her hay, but with the kind she likes, fed in moderate amounts and placed in a properly constructed rack she does a pretty good job of cleanly eating it. Any that she tosses to the floor can be gathered up, if the floor is clean, and placed in a rack outside in her yard to be picked over again at her leisure and the residue set aside for bedding.

Goats are very active animals and need as much freedom as you can possibly allow them. If you have sufficient space, the happiest arrangement for your animals is a fenced-in pasture. This permits them to roam about selecting the growing things that appeal to their appetite, gives them needed exercise and opportunity to find a shady spot when the day is hot, and protects them from dogs.

It is well, however, before the goats are given freedom of the pasture, to make a careful inspection for poisonous plants. The number of these to be found throughout the country runs into the hundreds, but they are not all found in the same locality. There are certain species rather widely distributed among which are the various types of laurel and these you should learn to recognize, for they are deadly poisonous—sometimes in very small amounts (I have known of a goat dying after eating a little of a discarded Christmas wreath). Wild cherry also is widely scattered and the early shoots which young stock is apt to eat, and the leaves in their wilted state will produce prussic acid poisoning, frequently with fatal results. Goats can eat the more mature leaves and the dried wild cherry as well as the bark with apparently no ill effects. Roots of the water hemlock are most deadly to all animals, but since goats do not burrow for their food, danger to them from this plant is less frequently met. Common bracken, of the fern family, is another plant to be avoided. Loco weed is a very dangerous plant of which there are many varieties and oddly enough it is of the same botanical family as the much desired alfalfa.

These plants are not generally found in open pastures, but in the wooded and swampy sections. Your state department of agriculture will give you information about the poisonous plants native to your particular locality and it is your job to clear them out if they are present in your pasture.

The usual signs of forage poisoning are vomiting, frothing, staggering and convulsions with cries of pain and erratic leaping about. Sometimes, if the attack is not severe a teaspoonful of bicarbonate of soda placed dry into the animal's mouth will increase the vomiting and help to throw off the poisonous cud,

**Five Poisonous Plants**

*Upper left.* Brake fern. *Upper right.* Loco weed, *Aragallus lambertii. Center.* Mountain laurel. *Lower left.* European Hemlock. *Lower right.* Wild cherry.

or epsom salts as a drench will purge it from the system. Syringes for administering liquids may be obtained from veterinary supply houses, and it is well to have one on hand, but in emergency a tonic or similar bottle may be used.

Drenching a goat, however, is a very serious undertaking and should not be resorted to unless no veterinarian is available. Under no circumstances should a novice or nervous person try it. If it must be done, get the animal onto her feet and put one leg on either side of her body holding her with pressure of your knees. It may be necessary to tie her. Insert the neck of the bottle or syringe at the side of her mouth, but don't press down on her tongue. She must have free movement of her tongue in order to swallow. Don't raise her head above a horizontal position and be very careful to pour the liquid slowly and gradually in very small amounts. If she struggles, stop. It's better to spill the liquid than to get it into her lungs, for that will mean her death. Follow the treatment with warm goat milk or black coffee, using the same care in administering it. Be gentle and talk to her so that she will not be frightened. In treating a very sick animal or a kid to whom you hesitate to give a drench, a wad of absorbent cotton soaked in the liquid and pressed between the goat's teeth at the side of the mouth, although a slow process, is effective.

If it is necessary to tether your goats, use a chain or hemp rope, not too heavy, with a swivel at either end to prevent tangling. Never use cotton, clothesline rope. Change their grazing place each day (they will not eat on ground that has been soiled with droppings) and place water within reach. Check on them occasionally to be sure that they are safe and comfortable and are finding sufficient green food to keep them contented.

In any case have at least a small, fenced exercise yard just outside their quarters so arranged that they can go inside for shelter from storm or sun. The yard is best on the south side so that it can be used even in winter when it would be impossible to tether them out-of-doors. This little yard will save you

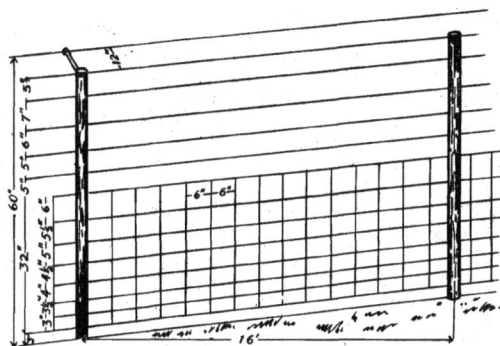

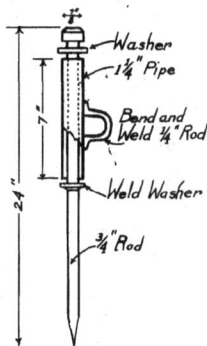

A post and wire fence for goats and a non-tangle tethering stake almost any garage man can easily make.

much anxiety, but it is an exercise yard solely and not meant for grazing.

If you value your goats don't tether them out and go off for the day. You are apt to return and find that a goat has been attacked by dogs or has tangled in her chain and strangled. These two accidents are the most frequently reported disasters that happen to goats. In this connection it is not advisable to encourage a too friendly relationship between your dog and the goats. Goats and dogs can and do get along amicably, and work dogs are used on the ranges for herding sheep and goats, but with the household dog living in close, intimate contact with the household goat, there is always the danger that an incident will arise when the goat will butt the dog who will retaliate in a dog's language, and even a powerful buck is not equipped by nature for prolonged battle with an angry dog. My own dog, brought up with the goats from puppy days, helped to kill a goat whom apparently she loved above all the rest, and from various goat owners I have known many cases of goats killed or badly torn by dogs.

A four- or five-foot cattle wire fence makes a very substantial enclosure and protection for goats, or a two-wire electric fence is used by many people with success, but in no case should barbed wire be used. With a high fence it is not needed, and a low fence topped with barbed wire will not restrain a goat nor keep out a dog, but will mean instead ears or udder torn by the wire. If you cannot have a fenced-in pasture and must tether the goats, whenever possible take them with you on a trip through the woods. They will follow along as happily as a dog and stay closer by your side. Or while you are working out-of-doors let them wander about, but watch out that they don't nip off your choicest roses. If you fear for your trees, for as the season wanes some goats will nibble at the bark, smear the trunks now and then with a broom dipped in manure and they will keep away.

On days when the goats cannot be staked out it is well to cut green feed or to supply them with succulent feed such as beet pulp soaked in water, or vegetables—carrots, turnips, cabbage—almost any kind of vegetable except onions which they don't like. A general rule as to quantity to feed is 6 pounds per 100 pounds of body weight. An average doe might be fed 4 pounds of green stuff twice a day. This will keep up their milk production.

In fact, it is a good plan to plant a few rows of root vegetables and cabbage in the kitchen garden just for the goats. The English goatkeepers go in strongly for kale which is hardy and grows late into the fall. At all times cuttings from vegetables and fruit are good goat food and help to produce milk. If you add such food as a new part of a goat's diet, you will very likely see a definite increase in milk production. Such succulent food should form part of the goat's diet especially when pasture is limited or the grazing season ended.

Grain is a concentrated food and should be fed with caution. Too heavy grain feeding may result in indigestion or overproduction of milk with ultimate burning out of the goat's ability to produce any milk at all.

In the early days of goatkeeping, the goat owner had to mix his own feeds—a somewhat hit or miss procedure even when a specific formula is followed. There is no way the layman knows for determining how much actual food value there may be in a given bag of oats or corn. And mixing your own feed is a real chore. Your spirits droop when you see it getting low in the barrel. Nowadays, the various grain companies prepare grain combinations especially intended for goats, feeds which contain all the essential components, including minerals, which perhaps the individual might find unobtainable. For example, people living in New England, Delaware, Pennsylvania, and the Eastern Shore of Maryland can obtain a sheep and goat feed manufactured by the Eastern States Farmers Exchange (a cooperative association) which goats will eat day after day with relish. The formula for this food, developed after actual experiments with goats, lists yellow corn, crimped oats, bran, linseed and soybean meal, corn gluten meal, molasses, iodized salt, irradiated yeast, calcium carbonate, dicalcium phosphate (found in milk). Now what individual can attempt to assemble an assortment such as this for his goats? Yet these ingredients are, in the judgment of competent people, essential in a balanced diet.

Wherever you live, you will probably find that the local grain company carries a feed prepared specifically for goats and unless you raise your own grain in properly fertilized soil it is easiest and safest to rely on the product of the grain dealer. Dairy rations may be had in finely ground form, but most goats prefer the coarser mixtures or the pellets—something that they can crunch.

The amount of grain fed depends on the needs of the individual animal.

For milkers it is determined by the milk production, the ratio being 1 pound of grain with a protein content of 16 per cent for each 3 pounds of milk (approximately 3 pints) the goat produces.

For pregnant does: In the diet of pregnant does, the grain feeding should be lessened with their lessening milk production, and during the last two months of pregnancy when the doe should be dry it should not exceed 1 to 1½ pounds daily. To this may be added bran for bulk and laxative quality.

For dry stock the allowance is 1 pound daily which may be increased if the animal is thin and in need of additional food.

For bucks 1 to 2 pounds daily is adequate, depending on the size of the animal and the service required of him.

For kids: Kids should be offered grain at one to two weeks old—a fine mixture such as a calf starter at first—and at about a month old they too can be given the coarser grain mixture fed to the older goats, as much as they will eat up at a feeding.

Grain should be given in two feedings, usually before milking. The doe is then contented and quiet while being milked. But whether you feed before or after milking be consistent in the method you use and always clear away any remnants of feed or hay in the racks before putting in the new meal. Sometimes people have complained that their goats don't eat up their grain, and on inquiry I have found that the fresh supply is put in on top of the leftover. Goats like a clean table to eat from just as humans do. Perhaps the goat has rested her hoof on the edge of the feed box opening and manure has dropped in—not enough to be apparent except to her sensitive nose—and in this event she will refuse her grain no matter how hungry she may be.

Some people fix a container on the milking bench and feed the goat while she is being milked. I tried this method when I had only a few goats, but it meant waiting until the doe finished her grain even though the milking was completed, or forcing her to leave the unfinished dish reluctantly. Also I found that some goats objected to eating from another's pan—one doe would not eat her grain unless a clean paper was spread on the shelf! Fed in their own stalls, they finish their grain down to the last speck and none is wasted.

If a goat goes "off her feed" as occasionally they do, a meal of bran or stale bread cut into small pieces will usually restore her appetite, or plain oats or some delicacy such as corn

142

When grain is bought in small quantities, metal barrels are ideal for storing grain so rats can't get into it.

flakes. Experiment with her yourself until you find the thing that appeals. Sometimes putting such a goat beside a good eater will stimulate her appetite—the dog-in-the-manger idea.

Fed a properly balanced grain mixture supplemental feeding of minerals should not be necessary, particularly if alfalfa hay forms a part of the diet. But salt, preferably iodized salt which may be obtained in brick form from your grain dealer, should always be accessible.

### SILAGE FOR YOUR GOATS

Much as you will love your goats there are parts of the grounds where they can't be allowed to wander at will without damage to shrubs and trees and to the garden. These are the places where your lawn mower and sickle get into action to keep weeds and grass under control. If the goats are getting plenty of pasture these cuttings won't interest them much, if at all, even though they would like the chance to make a raid on them on their own. Yet if a spell of dry weather comes, or several rainy days that mean housing in the barn with dry roughage, these very clippings would be accepted readily. If you are handy, a small silo may be constructed into which the clippings may be stored to be fed out at just such times. Any forage that the animals like to eat green will make suitable silage, just take care that no wild cherry, which sometimes sends out new shoots close to the ground, laurel or rhododendron leaves, or other poisonous plants are included in your cuttings.

There are a few fundamentals to be kept in mind in the making of such silage. It is most important that air be excluded and to this end the sides of the silo must be made airtight and the material packed in tightly, either by tramping it down if the silo is large enough or tamping it and placing a weight on top, being sure that it is distributed evenly so that no air spaces are left at the sides, for the presence of air causes oxidation and the silaged material will get moldy and unfit for use. The clippings

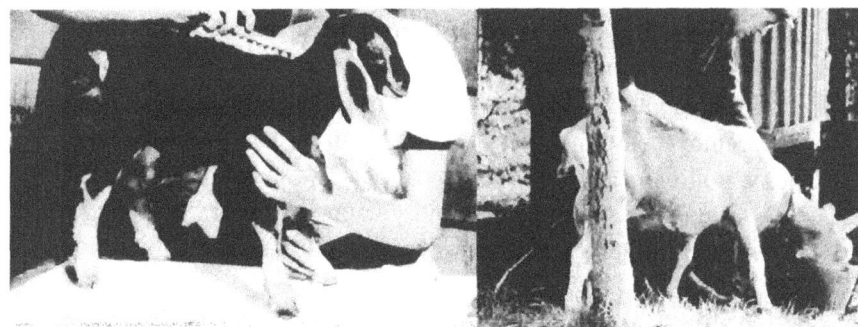

*Left.* An ordinary 10-cent scrub brush makes a good brush for goats. The people are Ed and Carolyn Robinson, authors of "The 'Have-More' Plan." *Right.* A small silo made by the local blacksmith.

are best chopped into ¼-inch lengths and in wilted condition when placed in the silo. Technically, the proper moisture content when put in is 68 per cent—not wetter, and not drier than 58 per cent. The rapidity with which the cuttings will dry on the ground depends, of course, on the size of the swaths, the amount of sunlight they get, the humidity of the air, and the wind. If they are too dry when added, packing is difficult and some of the valuable carotene is lost, and if too wet much of the juice will leak out, be smelly and attract flies, and eventually rot the silo. Also, moist clippings make less palatable silage with less food value. If the material seems too wet and it isn't advisable to leave it longer on the ground because of the possibility of rain or because of other duties, finely chopped, good quality hay may be added which will soak up some of the moisture.

When the silo is full its contents should be leveled off and pressed down thoroughly and again in two days' time pressed down again and any spaces at the sides filled to exclude air. This pressing should be continued once a week until the material has settled evenly. The heaviest and greenest material, which dries more slowly, should be placed on top.

In removing for feeding, take pains to keep the material level and remove any that has spoiled, to prevent further rotting. Some will be messy on top. After opening for feeding it is well to continue to use the silage rapidly enough to prevent spoilage, as exposure to air each time will cause the material to deteriorate.

A small silo can be made from a molasses drum obtainable from a bakery. A tight-fitting cover can be made from plywood —and adhesive tape may be used to fasten this to the drum and so seal silage from the air.

If your small silo is a success and the goats like the food— some will eat it, others perhaps refuse it—you may want to go into the making of it in quantities large enough to help out considerably with winter feeding and to provide yourself with more elaborate equipment. You can learn a lot about the making of silage from *Leaflet BDIM–Inf–38, USDA*, "Making Grass Silage by the Wilting Method," and *Farmers Bulletin No. 578 USDA*, "The Making and Feeding of Silage."

Another way of providing "silage" for your goats during the winter is simply to plant a small plot of sugar beets, mangels, or carrots. These may be stored in a root cellar. Goats will eat them with relish if they are washed and sliced.

Most small goatkeepers do not bother feeding silage. However, silage in the winter can mean more milk. If you get on the right side of your grocer he will save you lettuce leaves, cabbage leaves and carrot tops. Sometime in the near future it will be possible to buy canned silage from your feed man.

# FEEDING THE MILKING GOAT

**Bulky Foods.** Goats are *ruminants*, and their specially designed digestive system makes it possible for them to *chew the cud* and to digest *bulky, fibrous* food. Hay and hedge trimmings are examples of foods that are bulky and that contain the tough food material called *fibre*.

Grass, mangels, cabbages, kale, and the many green plants that grow by the roadside are other examples of bulky foods, but these are bulky because they contain a large amount of *water*.

The fibrous foods give work for the digestive organs and for the teeth, and this helps to keep the goat healthy. The watery foods have a good effect upon the bowels : that is to say, they help to keep the food passing regularly through the animal.

Hay is made from grass and clover, from lucerne or sainfoin, and from the many wild green plants that grow on waste land. Some goat keepers feed hay only during the autumn and winter, but others feed it all the year round, giving their goats a little before turning them out to graze in the morning. This helps to prevent the *scouring* (see page 44) that is caused by eating *lush* (juicy) or dew-covered plants.

An animal eating grass in a grass field is said to be *grazing*. A goat nipping off the heads of weeds and the shoots of bushes and trees, taking a mouthful here and a mouthful there, is *browsing*. Although goats will eat grass, they prefer browsing to grazing. The best land for goats is, therefore, that on which many wild plants and bushes grow. Even brambles and gorse (furze) are liked. The leaves and shoots of elm, ash, hazel, willow and holly are among those that can be fed. When the goats are stall-fed these wild plants have to be collected and brought in. Feed as much as each goat will readily clear up. When a goat has had enough she will lie down and start cudding. Any food left over should be removed before more is given.

A poisonous plant is one that upsets the goats even if it does not kill them, and a large number of plants are suspected of being poisonous. It is not, however, true to say that *all* these are *always* poisonous to *all* goats. Some plants seem to be more harmful at certain stages of their growth or at certain times of the year. Goats herded or on free range are probably less *susceptible to* (likely to be hurt by) poisons than those whose diet is restricted by tethering or stall-feeding. Some goats are more sensitive to poisons than are others, and the danger seems to be greater when the goat has an empty stomach. It is also possible that for every poisonous plant there is an *antidote* that neutralizes the poison, and that goats on free range find these plants that are antidotes. Thus it can be seen that there is a great deal more to be learnt.

However, the following plants have so frequently been the cause of trouble that they should be avoided altogether :

| | |
|---|---|
| Yew. | Laburnum, especially the pods and seeds. |
| Rhododendron. | Ivy berries. |
| Laurel. | Ragwort. |
| Box. | Wild clematis. |

A great deal of valuable, bulky food can be grown in the garden (see Picture No. 18), and sugar beet pulp, soaked, is useful too.

**Concentrated Foods,** e.g. oats, flaked maize, bran, linseed cake, etc.

These contain neither a great deal of fibre nor much water. Therefore, they are not bulky. Their value lies chiefly in the *starch* and *protein* that they contain. Some of them are well balanced as regards these two food materials, whereas others are richer in either starch or protein. These concentrates can be mixed together in such a way that they are suited to the production of milk. The mixture is then *balanced for milk*, and the amount fed should be in proportion to the amount of milk given by each goat. The yields must, therefore, be measured. See Picture No. 26.

There are many satisfactory mixtures of concentrates, and many of these are described in other books. When concentrates are plentiful the amount fed is usually 5 ounces for each pound of milk yielded. It is usually fed in two feeds, one at each milking. Concentrates are used most economically when combined to form a balanced ration and when fed in proportion to the milk yields. Problems of feeding cannot, however, be always solved by mathematics, and the calculations must be used as a guide and not as laws to be rigidly obeyed. See page 26.

Ready-made mixtures, called *compound* cakes and meals, can be bought. Those sold by reliable firms are quite satisfactory. The mixtures are sometimes made up in the form of cubes, called *nuts*.

**Maintenance and Production.** An animal's food supplies energy, heat, and materials for the repair of the body. This is called *maintenance*, since the food maintains the animal alive. Food is also used to produce milk, growth, or the birth of a new animal. This is *production*. It is convenient to think of a ration as in two parts (maintenance and production), although the animal itself makes no such distinction.

In summer when there is good grazing and a little first quality hay, these two together may provide enough for maintenance and the production of, say, 2 pints of milk. Concentrates need then only be fed to those goats giving more than this.

In winter, when green food may be scarce, the maintenance ration for a goat of average weight (100-150 lb.) might consist of 3 lb. of hay, 2 lb. of roots and ½ lb. of concentrates per day. Milking goats would, in addition, receive their production rations according to their yield.

Between these two extremes there are many satisfactory rations based upon the amount and quality of the green food and roots, the amount and quality of the hay, and the amount of concentrates available. Concentrates can be economized by feeding home-grown green crops, and the most valuable are those that are most rich in protein, e.g. lucerne, sainfoin, vetches and clover.

**Scraps and Economy.** Stale bread, dried in the oven, can take the place of some of the starchy concentrates in the rations of any of the goats. The older goats can be given carrot tops, the outer leaves of cabbages and cauliflowers, pea and bean haulm, and the stalks of Brussels sprouts (split lengthwise), and *cooked* potato skins. The cooking can be done in the oven so that the skins are crisp. Goats are very particular about the cleanliness of their food, so the scraps must be kept clean.

Some people feed acorns and chestnuts to their adult goats, but the amount must not exceed about 4 ounces per goat per day. They must be introduced gradually into the diet, and must be quite ripe and, preferably, dried.

Surplus vegetables from the garden can always be fed, and the only ones that must be cooked are the potatoes.

Goats are very fond of dried nettles, and some people find that dandelion leaves tend to increase the milk yield.

**Minerals and Vitamins.** In addition to protein, starch and fibre, animals need fat and sugar in their diet. There will usually be enough of these in all normal rations. Animals,

18. The goat plot, which must be fenced. Lucerne grows on the right and Perpetual Kale in the centre. Other useful crops are mangels, swedes, turnips, carrots, beetroot, cabbages, marrow-stem kale, thousand-headed kale, maize, vetches, clover, chicory and sainfoin. If the soil is sour it should be limed, for this benefits the goats by making the crops healthier and more nourishing. The building in the background is a barn and fodder store.

19. Some people feed hay in nets. The amount fed daily depends upon the size and condition of the goat, the quantity of other foods available, and the weather. An average quantity consumed by an adult goat in a year is 8 to 10 cwt. Unlike other animals, goats will thrive on hay that is made from the plants that grow wild by the roadside and on waste land.

especially those giving milk or growing rapidly, also need minerals and vitamins.

The two most important minerals are lime and salt. If the goats are having green food, a little clover hay, and a lump of rock salt to lick, and are out in the sunlight whenever possible, they are unlikely to suffer from a shortage of minerals. Some people, however, fix *mineral licks* in the goat house and add salt to the goat's drinking water. The mineral licks are probably a necessity when the goats live in a district in which the soil is *sour* (i.e. lacks lime).

Sunshine and green food are safeguards against a shortage of vitamins. During the winter, however, a heavy milking goat and the growing kids may benefit by the addition of a little cod-liver oil to their diet.

**Water.** Fresh water, and plenty of it, is as necessary as good food. A milking goat may drink as much as 2 gallons a day, though the more green food that she eats the less water will she need.

**General Rules of Feeding.** Although the food itself is important, the methods of feeding are even more so. The main rules of good management are these :

1. Meal times must be regular. Two feeds a day are enough for the adult goats and goatlings, though when stall-fed they benefit by an additional mid-day feed of green food.

2. All food must be fresh and clean.

3. The troughs, mangers and buckets must be kept clean. Any food left over from the last meal should be removed.

4. Any big change in rations should be made gradually, or the goats' digestions will be upset.

5. At meal times the goats should be watched. An animal that is not eager for its food is out of condition. The state of the droppings is a good indication of the state of the animal's digestion. See page 41.

6. Each animal's likes and dislikes must be considered, and the amount fed to each must be varied to suit its condition. It is easier to maintain condition and milk yields than to regain them once they have been lost. For example, rather generous feeding towards the end of the summer will help to prevent the drop in the milk yields that usually comes in the early autumn.

There is still much to be discovered about the feeding of animals. Although based upon certain general rules it is an art than can only be learnt by experience. The more we learn and the more we study our goats the better stockmen shall we become, and the more critical shall we be when we read persuasive advertisements,

## Feeding Your Goat.

You may consider your goat primarily as a wartime factory for the conversion of waste produce into milk. It is not *quite* as simple as that, though it is a captivating thought. It depends on the waste.

Contrary to comic paper conception, a healthy goat will not eat *anything*. A starved goat will; so will a starved human.

The goat's general cleanliness of habit extends to its feeding which further enhances its value as a dairy animal. Above all things, a goat desires clean food, and does not like other animals' or people's leavings.

If I give my goat a quartered apple she will eat it with relish. If I take a bite out of an apple and hand her the remainder, she will refuse it.

Perhaps I spoil my goats (though I think that pays), for I always carry a few toffees or biscuits in my pocket when I visit them, and they will nuzzle me and my pocket until I produce one. Should that toffee drop from the goat's mouth to the floor, however, she will not touch it again. I have tried disguising the toffee in a fresh paper, but even then it is refused. If hay, pulled from the rack, is dropped, it is never eaten.

A goat may eat and drink from the same buckets as her own particular stable companion, but never after any other animal.

These points are made to press home the type of animal with which you are dealing, and I repeat, they are definitely in the goat's favour. One can really fancy the milk from such a dainty creature.

There are scientific methods of feeding goats for high yields, and there are rules for balancing rations so that daily the goats receive their correct proportions of proteins, carbohydrates, vitamins, and the like. I do not propose to touch on this aspect of feeding, for such information can be obtained from many authoritative works. Besides, the war has complicated this subject, and many and strange

*A Kodak Snapshot.*

It is a popular fallacy that goats will eat anything. True, they are fond of tit-bits and have a sweet tooth, but in every respect they are dainty and fastidious over their food, and will never touch anything that is soiled. You can really fancy the milk from such a clean animal.

are the foods and mixtures now fed to goats which would have given rise to horror before the war. Yet the results are found to be extremely good !

Rationing of animal feeding stuffs is also in force now and places further restriction on what we can and cannot feed to our goats. Goat-owners are, however, much more fortunately placed than other small stock keepers—in respect of feeding stuffs goats have been allowed, by the Ministry of Food, the same priority as is enjoyed by dairy cows. In order to obtain feeding stuffs the new goat owner must register with his County War Agricultural Executive Committee, the address of which is the county town.

I shall tell you of the various corn foods (concentrates), green foods, roots, etc., you can feed to your family goat, how to estimate her requirements, and so on. But do not, I beg of you, follow me by rule of thumb. Study the individuality of each goat not only as to her food fads and fancies, but also as to her appetite. Try to feed as much home-grown food as possible. Also, do not be afraid to experiment with whatever foods are available in your district. Perhaps, for instance, you will be in a position to obtain brewery " screenings " which, soaked, can be added to mixtures in place of oats, with excellent results.

The Champion British Saanen which broke the world's officially recorded milk yield was fed on a simple concentrate ration consisting of 3 parts flaked maize, 2 parts broad bran and 1 part crushed oats, by measure. Of course, in addition she had tit-bits such as baked bread, and hay and greenfood. On this ration she produced an average approximating to 1½ gallons of milk daily for 365 days.

And why I say don't be afraid to experiment is because I know of a cottager's scrub goat which, on such simple feeding as grass, cabbage leaves, and yellow meal boiled with wheatings in a gruel, has given 6 to 7 pints of milk daily.

As with dairy cattle, so with dairy goats, feeding is considered in two parts—maintenance ration and production ration. A normal goat of approximately 135-140 lbs. weight will maintain herself in health and condition and yield two or three pints of milk daily on what she can pick up for herself during the summer on good browsing or on a daily menu of 3 lbs. hay, 3 lbs. cabbage or roots

154

| FOOD. | MILKERS, IN-KIDDERS, AND GOATLINGS. Parts by Weight (read each column down for complete ration). | | | | | | | | | | | | | | | | | MALES. | KIDS (from one month old). |
|---|---|---|---|---|---|---|---|---|---|---|---|---|---|---|---|---|---|---|---|
| Bean Meal | | | | | | 1 | 1 | 1 | 1 | | | | | | | | 3 | 2 | |
| Beans, Kibbled | | | | | 5 | 5 | 3 | | | | | 1 | | 3 | 3 | 3 | 1 | | |
| Beet Pulp (Dry) | | 2 | 1 | 3 | 3 | 3 | 2 | 2 | 3 | 3 | 2 | 2 | 3 | 2 | 2 | 5 | | | |
| Bran | 1 | 1 | 1 | | | | 1 | 1 | 1 | 2 | 2 | 2 | | | | | | | 1 |
| Bread (stale) | 2 | | | 2 | | | | | | | | | | 2 | | | | | |
| Cotton Cake (decorticated) | | | | | 2 | | | | | | | | | | | | | | |
| Cottonseed Meal | | | | | | | | | | | | | | | | | | | |
| Dairy Nuts | | | | | | | | | | | | | | | | | | | |
| Dried Grains | 5 | | 3 | | | | 1 | 1 | 2 | 2 | 3 | 3 | 2 | 3 | | | | | |
| Earthnut Cake (decorticated) | | | | | | | | | | | | | | | | | | | |
| Lamb Food | 5 | | | | | | | | | | | | | | | | | | |
| Linseed Cake | 1 | 7 | 2 | | 4 | 9 | | | 1 | 4 | 1 | | | | | | | | 2 |
| Maize (flaked) | 1 | 6 | 2 | | | | 1 | 1 | 1 | 1 | 3 | | 4 | | | | | | 1 |
| Maize Gluten Feed | 1 | | 1 | 1 | | 3 | | | | | | | | | | | | | |
| Maize Meal | 1 | 1 | | 1 | | | | | | | | | | | | | 4 | 1 | |
| Middlings | | 1 | 2 | 1 | | | | | | | | | | | | | | | |
| Oats (crushed) | 1 | 1 | | 1 | | | | | | | | | | | | | | 1 | 2 } (crushed fine) |
| Roots (pulped) | 1 | | | | | | | | | | | | | | | | | | 1 } |
| Soya Bean Meal | | | | 2 | | | | | | | | | | | | | | 1 | |
| Soya Cake | | | | | | | | | | | | | | | | | | | |
| Wheat (flaked) | 1 | | | | | | | | | | | | 2 | | | | | | 1 |

(or their equivalent), and 1-1½ lbs. of concentrates (corn or meal foods).

To maintain a yield of milk over 2-3 pints, the goat must have an additional production ration of approximately 6 ozs. of concentrates per lb. (roughly a pint) of milk. Thus if your goat gave six pints of milk you would give her roughly 2½ lbs. of concentrates daily. But experience will teach you that the actual amount depends upon individual appetite and capacity for conversion of food into milk, and upon the kinds of other foods available.

The appended rations chart will show you how to make up complete menus for your goats, according to the foods you can obtain in your district.

The complete concentrate ration should be divided into two parts, one fed in the morning and the other in the evening, at strictly adhered to times. You will soon learn how many tinfuls or mugfuls of certain meals make a lb., so that there will not be the need to weigh out the meals separately each day. They should be fed dry as a rule.

Variety in greenfood feeding is one of the secrets of keeping goats well and contented. And there is plenty of variety to be obtained even in one's own garden or neighbourhood—cabbage, kale, prickly comfrey, chicory, tares, artichoke tops, pea haulm, sunflower foliage, beetroot and turnip tops, grass clippings from the lawn, sheep's parsley, cow mumble, docks, nettles and other common weeds, fruit and ornamental tree prunings, hedge prunings, rose bush prunings, gorse, heather, brambles, briars, nut, and so on.

Evergreen trees, hedges and shrubs should be avoided, however, as the majority are poisonous to goats. Three exceptions are the evergreen honeysuckle, *lonicera nitida*, a quick-growing hedging plant of which the goats are fond, holly, and ivy without the berries.

In winter, when greens apart from hardy kales are scarce, the goats can have such roots, sliced or pulped, as kohl rabi, carrots, mangolds, turnips, and cooked potatoes and artichokes.

Greens and roots should be fed at midday, and changes may be rung by pulping them and "drying them off" with bran or middlings, to eke out the concentrates.

Hay is an essential, and in my opinion should always be before the goats. There is nothing like good, sweet hay for producing milk. If you cannot make your own hay, it can be bought by the truss, costing from 2/6 to

4/6, and a truss will last a goat about a month. I have frequently brought a truss of hay home on the back of my old car from some outlying farm.

There is good and bad hay. The best *is* best—it will produce more milk, and less of it will be wasted. In order of preference, goats like sanfoin, lucerne, nettle, clover, and meadow hay.

Goats can receive their "minerals" via a block which they lick (and obtainable from any veterinary chemist's), or proprietary powder minerals which you mix with the food. You can also make up your own mineral mixture as follows: Sterilised bone flour, 5 lb.; finely-ground limestone, 5 lb.; common salt, 2½ lb.; and potassium iodide, 2 ozs. per 100 lb. of mixture—dissolved in a little water and sprayed over the mixture from a scent spray. 6 oz. of this mineral mixture should be used per 12½ lb. of concentrate mixture.

Goats must also have water. Milk contains 86.57 per cent. of water. Therefore plenty of water is the first essential in milk production.

The poor yields given by many goats is most frequently because either they do not receive sufficient water, or their "appetite" for water has not been developed. The latter is a point all kid rearers are advised to remember. All high yielding goats have a surprisingly large thirst.

Goats vary considerably in their reaction to water. Their fads and fancies must be pandered to, to induce them to drink more.

Goat keepers must remember that no goat will drink dirty water. Frequently she will not drink after another goat from the same bucket. Water that has been left standing in the goat house and is tainted with the ammonia from the litter will be refused.

I had a goat which would not drink cold water even on the hottest day of summer. But she would never refuse warm water at any time.

When goats change hands, change of water often upsets their desire for it. Used to soft water, they object to the new supply of hard water, or *vice versa*.

Water requirements need to be studied along with the diet. Thus a goat on a hay and concentrate diet requires and will drink more water than one receiving ample succulent green food.

Goats can, and should, be trained to a watering routine

from an early age. Water should be offered them at set times every day throughout the year, until "watering" becomes a fixed habit. It is of little use leaving water always before the animals.

Finally, when a goat refuses water, the best inducement is to make the water warm, and drop in a little salt or middlings.

## THE ROUTINE OF STALL-FEEDING.

The following is an account, specially written for me by Miss N. M. B. Smales, of how a goat-owner manages her small herd on the semi-confined, or stall-feeding, system:

Much has been written on the different ways of keeping goats, but I hope this will be of help to amateurs who are wanting to start with goats but are afraid they have not sufficient space for them. I have always kept my goats on the "stall-fed" system and have so far managed to keep them fit and well. Never have I had more than a very small patch of ground for their run or exercise yard.

All this last summer and winter I have had three milkers and reared three pedigree kids, two from the age of three weeks and one who came to me when she was about six weeks. The goat house is small and comprises three stalls and a loose box. The hay rack extends along the length of the stalls, and there is a similar one, only built nearer the ground to enable the kids to reach it, in the loose box.

The goats lie on creosoted boards which are only slightly slatted, being made from broken up bulb boxes, one to each stall and two for the loose box to facilitate handling.

The house, 10ft. x 6ft., is standing at the back of an enclosed space about 20ft. x 20ft., which is the only space available for the exercise yard.

The following details of daily routine, making allowances for weather conditions, etc., will show you exactly how my stall-fed goats are managed.

*Summer routine—*

7.30 a.m. Each goat receives half her daily ration of concentrates according to her personal needs, and her milk yield.

Milking starts immediately after all are fed.

8.30 a.m. If the weather is fine, every animal is turned out into the run. Branches of leaves are hung up for

them or cut grass or other fodder is put into their racks.
Cleaning out is done when convenient.

12-12.30. If it is very hot and the run is not shaded,
all are put into the shed, green food is put in the racks,
and the goats are left until it is cooler. I believe they
are far better lying quietly in their stalls contentedly
chewing the cud and making milk, than trying to get out
of the way of flies, or panting in the heat.

Have you ever noticed that a herd of cows out at
pasture will nearly all be lying down during the middle
hours of the day? It is while chewing the cud that the
milk is mostly formed.

Of course, if the weather is typically British and there
is no very hot sun, the goats can remain out in the run
all day, as long as the shed door is left open for them
to go in and out as they wish.

6 p.m. Milking time, when each animal receives her
last ration of concentrates. As soon as they have finished
this I usually take them all out for a walk in the lanes
and droves, where they can eat and browse to their hearts'
content in the cool of the day. We are generally out
for 1 to 1½ hours.

Upon their return the goats are put to bed, and if I
do not think they have had enough, I " top up " with hay
or green food.

*Winter routine—*

8 a.m. First feed of concentrates, and milking.

9 a.m. Racks are filled with hay or ivy.

10 a.m., or as soon as it is warm enough, the goats
are turned out into the run.

Cleaning out takes place when convenient.

12-12.30. The goats are put into the shed and given
cabbage leaves, roots, or vegetable parings, etc., when
available. If not, a small feed of hay. If the weather
is suitable they are turned out again about 2 p.m., but
if it is at all cold or dull they are kept in.

5.30 p.m. The evening feed of concentrates, and
milking, followed by a full rack of hay.

At about 7.30 I go out and give them all a really warm,
mealy drink, which they appreciate very much. This is
made by pouring boiling water on to a double handful

of weatings, or bran. Sometimes I add salt or black cattle treacle. Cold water is poured on just before I take it out to them.

In the winter, daily exercise is taken in the afternoons when possible—a brisk walk on the roads. At the time of writing the Foot-and-Mouth restrictions are in force here, and I have been unable to take my herd on to the roads for the past month. So the only exercise they have is a " scampede " once a day in a small patch of ground adjoining the house. The goats are remarkably well, and always seem happy and contented nevertheless.

Of course, this is only a broad outline of the management of stall-fed goats, and must not be taken as the only way. For instance, you may say: " It is all very fine, but I have to be away at work all day and can't give them a mid-day meal; nor can I put them in or let them out during the day."

All right! Make your own plan, by using your commonsense, and leaving the goats sufficient food in their racks to last them till you come home in the evening. As long as they are able to get into shelter from the heat, rain, and cold winds they will not hurt.

But a word of warning. Do make quite sure that the boss of the herd (and there is certain to be one) does not stand in the doorway and keep everyone else out! And again, it is always wiser to have an entirely hornless herd or an all horned one, for the ones with horns have a distinct advantage over the hornless ones, who may be kept away from food and shelter.

Then there is the fodder question—the branches I spoke about. If you have no time to go out and cut branches for yourself, it is worth giving a reliable boy or girl a few pence pocket money to get it for you, as long as he or she knows what is suitable.

Or again, if you have a fair-sized garden and can allocate a portion of it to be the " goats' garden " you can grow all manner of things for them. To mention a few: Carrots, mangolds, swedes for root crops, and all kinds of cabbages and kales; artichokes (both tops and roots), green maize, oats to cut green, besides all the things you normally grow for your own use—the outside leaves of broccoli, curly kale, savoys, etc. A patch or two of lucerne is invaluable.

Personally, I seldom go out for a walk without taking

a piece of rope, a clean, small sack, and a strong pair of secateurs. There are endless things you can pick in the hedgerows and at the side of the road, which can go into the sack, to say nothing about the apples left to rot, or the mangolds, swedes and turnips that have fallen from a badly loaded putt.

The secateurs will cut those lovely pieces of nut, ash and elm that are in the hedge, and these can be neatly tied into a bundle with the rope. Do use the secateurs whenever possible, and do not cut the hedges indiscriminately or you will get into the farmer's bad books for damaging his hedges.

On more than one occasion last year I cut quite small branches from a huge ash hedge, taking pieces here and there so that no one could say I was doing any damage. Imagine my disgust when some week or two later I again visited that same hedge, only to find that the hedger had been busy, had cut the whole thing down, and was gaily burning all the now dried up ash branches. I rescued what I could and came home wishing to goodness I had not been so conscientious.

So if you *can* find out what hedges a farmer intends to cut that year, he will probably be willing to let you have first cut.

If you have to leave your animals for any length of time during the day, do be careful there are no weak spots in the fencing, for they are almost sure to find it out while you are away.

Also see that there are no traps that a kid or goat can get her leg or head caught in.

If you are at all doubtful of these points, then I say emphatically: Leave the goats in their shed where you know they are safe. It is horrible to be some distance from home and the thought suddenly come to you that all may not be well at home.

To keep the kids amused an upturned box gives no end of pleasure as long as it is firm and will not tip over. Logs for them to bark are a great source of delight to young and old, and are good for them, too.

To sum up. Although it may appear at first sight to be a lot of trouble and that you are constantly doing odd jobs with the goats, under the above system a methodical person can so arrange the work that there is ample time for the usual household tasks, gardening, etc., and the goats' feeding times can be arranged to suit each individual household.

On the stall-fed system you know exactly what quantity of food each goat receives daily, and you are able to give that little bit " extra " to a particular animal which may need it, for no two goats can be treated exactly alike and careful observation soon tells you when each member of the herd is " doing " to the best of her ability.

Goats are always amenable and easily handled when stall-fed because they are as one of the household and treated accordingly.

One last point, when buying a goat which you intend to keep like this, it is advisable to buy one that is not used to a large field or " free range," because it may take a long time for her to settle down to the new way of living, and so cause disappointment to her new owner.

Experience was most costly upon one occasion—a goat was bought which had been running free wherever she willed, and, having horns, and being most self-willed, in no time had she charged clean through every wire netting fence she came to! Moreover, unused to high feeding, she could never digest corn feed and nearly " died on me " several times!

# FOOD

O N the mountains and in the valleys of the United States the Angora has had a variety of food. He is a natural browser, and will live almost entirely on brush, if this kind of food is to be found, but he readily adapts himself to circumstances, and will live and do well upon an exclusively grass diet. The fact that the goat is a browser has been made use of in clearing farms of brush and objectional weeds. If a sufficient number of goats are confined upon a limited area for a period of time, they will kill most of the brush upon this land. They will eat almost every kind of brush, but they have their preferences and enjoy especially blackberry vines and those kinds of brush which contain tannic acid, such as scrub oak. They do not poison easily, and if there is a variety of food they rarely eat enough of any kind of poisonous plant to prove fatal. If, however, they are hungry, and have access to places where there are poisonous plants, they will eat enough to kill themselves.

## KILLING BRUSH.

If one wishes to clear brush land, he should confine the goats to a comparatively small tract. The goats kill the shrubs by eating the leaves and by peeling the bark from the branches and trunks of the trees. The brush thus deprived of lungs, soon dies and the roots rot. As fast as the leaves grow they must be consumed, so it is well to allow the goats to eat most of the leaves off of a limited tract, and then in order to give the goats plenty of feed, they should be moved to another field. As soon as the leaves on the first tract have regrown the goats should be again confined to this land. In this way the leaves are continually destroyed. This process can be continued as fast as the leaves regrow. By this method it is estimated that a bunch of one hundred to one hundred and fifty goats will clear forty acres of thick brush in about two years. In countries where the grass grows as the brush dies, goats will eat some of this grass, but they prefer the browse.

On some of the older goat ranches, where the Angora has been raised exclusively for the mohair and mutton, it has become quite a problem to prevent the goats from killing out the brush. The goats have done well where other kinds of livestock would have starved, but as soon as the brush is killed the land produces almost nothing, and even the goats cannot make a living. To prevent as far as possible their killing the brush the flocks are moved frequently from one range to another, so that the shrubs have a chance to recuperate between visits. In this way brush can be kept almost indefinitely for the goats. On some of the western ranges, where cattle and sheep have, by continual cropping, killed much of the grass, good browse remains. These ranges would have to be abandoned if it were not for the goat. Goats do not in any way interfere with the pasturage of cattle or other livestock. Cattle feed contentedly on the same range with the goats, and this fact has led many southern cattle men to invest in goats. The goats are herded on the brushy lands, and the cattle range over the same territory and eat the grass. Horses have a great fondness for goats.

### SALT.

Goats, like other livestock, should have a small amount of salt. The salt should be kept where they can get it at liberty, or else it should be fed at regular intervals. If ground salt is given, care should be taken to see that individuals do not eat an oversupply of the salt.

### WATER.

While Angoras do not require as much water as sheep, yet they should be given a quantity sufficient at least once a day. In winter goats will live upon snow. Men have reported that their goats have gone for a week at a time, and all summer long, without any more moisture than they could get from browse and weeds, but even if Angoras should stand this treatment, they will thrive better with water once daily. It is estimated that under normal conditions a goat will consume about one-ninetieth of its body weight (about a pint of water for a grown animal) in a day. On hot days, when the animals are on dry feed, they will frequently drink two quarts of water.

## Kidding Without Worry.

To the beginner, I know well, the most to be feared part of goat-keeping is the first kidding experience. Anxiety increases at the approach of this time, and there is the fear that " everything will not be all right."

Such fears are absolutely groundless, as I and others have proved.

The following has been written for me by a breeder of very considerable experience, and I can vouch for every word of it: —

" When I started goat-keeping I knew very little indeed about goats. I started with an old non-pedigree nanny due to kid about a month after her arrival. I read all the literature on goats' kidding which I could get hold of, and the result was that as Pat's kidding approached I was simply terrified of what was going to happen, for the articles had told of all the complications which *might* occur so thoroughly that it seemed to me then that an easy, uncomplicated kidding was quite the exception.

" Pat kidded with no trouble at all, and in fact so easily that I was absolutely surprised. I have had many goats since old Pat's day, and have witnessed many, many kiddings, but I have only ever lost two goats through kidding, and both of these were old goats. Complicated kiddings I have found are decidedly the exception and I do not intend to deal with them here. The average healthy goat, not too highly bred, kids easily and is no bother. So, now feeling maybe, a little less worried over the coming event, what *can* you expect?

" The gestation period of goats seems to be about 150 days, but watch her carefully after the 147th day, for after then you may expect the kids at any time. Put her in a loose box, *untied*, with plenty of clean straw bedding. Don't leave food or water pails in the box in case of accidents. Watch her udder; if it is very large and tight don't be afraid to milk her before she kids, it will do no harm and may do a lot of good, but only loosen it; don't, of course, strip her.

" Towards the end of the gestation period the goat will look less bulky, her tail will seem to be carried higher, and her flanks will look hollow, and there will be a deep depression either side of her tail—in fact, the kids have dropped into the position from which they will be born.

" The first symptom is a thick-looking white discharge. The time elapsing between the start of this discharge and the actual birth of the kids varies considerably from an hour or two to several days, and so a watch must be kept carefully for further symptoms.

" The goat next becomes restless and will call out to you in quite a different tone from her usual bleat, a rather scared little noise in fact.

" The next stage is the change in the nature of the discharge. It will become a yellowish opaque one, rather like the white of an egg to look at. The kids are not far away now. By this time the goat will be scratching her bedding into heaps, lying down, getting up, remaking her bed and generally exhibiting every sign of uneasiness. She will generally go on like this for some little time until she lies down and starts to strain, slightly at first, but as her pains increase, more and more vigorously until the kid is born.

" As soon as the kid is born clean away the slime from its nostrils and with a finger wipe out the mucus from its mouth. The mother will quickly start to clean the rest of the kid and it will soon be on its feet looking for its first feed.

" If there are more kids to follow the goat will often start straining again even before she has finished cleaning the first one. Rub the kid down with a soft towel, then put him out of the way in a box of hay but where the mother can see it. The second, or second and third kids, as the case may be, will arrive quickly and with very little trouble. Clean their noses and mouths like the first one. All that remains now is the expulsion of the afterbirth. Generally this will follow quite quickly, but it is strongly advisable always to stay with the goat until you are *sure* the afterbirth has come completely away.

" The mother will be very ready now for a nice warm

bran mash and a drink of warm oatmeal with a dessert-spoonful of black treacle in it. The kids will be looking for their first feed; sometimes they are not clever enough to find the teats and just one lesson is needed. If the mother's udder is not sufficiently loosened by the kids, take a little out, but do not strip her for the first day or two until her full flow of milk has come in.

"Mother and family will now be ready to be left alone for a good rest and sleep, and so your worries are over, and you will face your goat's next kidding, I hope, with less anxiety."

## The Kids.

WHEN your kids arrive (never forgetting that twins and triplets are quite common) make an immediate examination of them to determine if they are male or female. If male, and unless fully pedigree on both parents' sides, do *not* keep them with the intention of breeding from them later. Either kill them at once, or keep them for fattening for table.

Immediately after birth they can be killed, like rabbits, by a sharp blow at the back of the head. Or they can be put to sleep at an Animal Centre.

But criminal offence as it now is to waste food, *don't* be squeamish about fattening kids for the table. Kid meat is a real delicacy, and the kids can be reared cheaply on milk substitute. More about goat meat will be found in another chapter.

You will keep nanny kids, of course, and you now have confronting you the " problem " of rearing them. In the old days they were just left with the mother, and you took what milk you could when the kids had finished.

Now we are wiser and produce better goats by rearing the kids by hand, and we save the milk for ourselves by switching the kids over after a time to a good milk substitute providing all growth essentials, yet costing very much less than the value of milk to us.

Hand-reared kids become real pets; you can train them to your own particular way of life, and they grow up into well-mannered, quiet adults who will answer your every beck and call.

It is certainly more troublesome to rear a kid by hand than to let it run with its dam—but here is where the family comes in. Your children will clamour to feed the kids, and my own little girl has proved that the young are perfectly capable of feeding the young.

Here are the essential facts on hand kid-rearing. Naturally, they may need modifying according to individual circumstances: —

It is important, right from the first, not to allow the kid

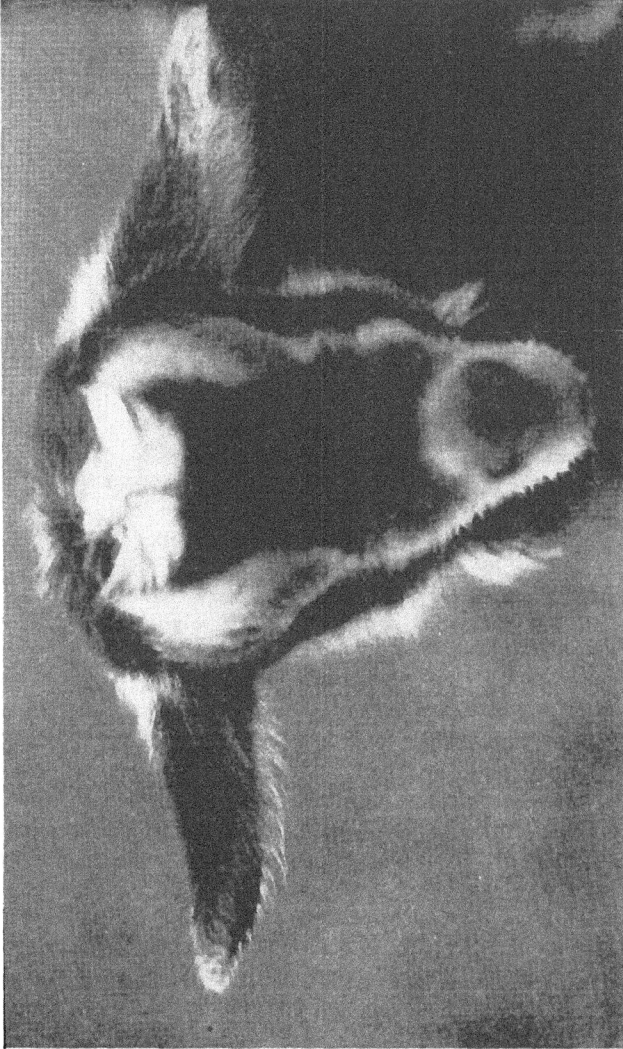

Full of charm, intelligence and high spirits, which they never lose as they grow up, who can resist the appeal of the kids to be loved and petted? The family goat-keeper has many happy moments in store when the kids come.

171

to suckle its mother; otherwise there will be great difficulty in parting them later, or in getting the kid to take
to a bottle. In some cases rearers have found it practicable
to allow the kid to remain with its mother for the first
two days, and then start it on its bottle-fed career.

The milk which first comes from the mother is thick,
yellowish colostrum—specially designed by nature for the
first few days of the kid's life. The kid must receive this,
but if for any reason it is not available, a half-teaspoonful
of liquid medicinal paraffin should be shaken up in the
milk or milk-substitute bottle once a day. The mother's
milk changes in character after two or three days, and
by the fourth day it is quite normal for consumption by
human beings.

For the first few days you can feed from an ordinary
baby's feeding bottle with teat and valve, 8oz. size. At
four days you will find a tomato sauce bottle, holding just
about ¾ pint, ideal. The neck should be straight, with
no ledges where milk can collect to go sour. At about
six weeks onwards a quart size vinegar or lemonade bottle
with a straight neck is excellent.

Following the baby bottle, use soft rubber lamb teats,
enlarging the hole with a hot steel knitting needle.

The milk must be fed at blood heat, *never* cold. It
can be heated in two ways: In a saucepan over the fire,
or by putting the milk into a bottle and placing it in a
jug of hot water until it is the right temperature. Test
the heat of the milk on the back of your hand.

Generally speaking, a strong, healthy kid will take about
8ozs. of milk each feed till about ten days or a fortnight
old, increasing to 12 ozs. Kids should never be forced
to take their bottles. If they refuse a bottle, or not quite
finish one, no anxiety need be felt unless it is a constant
occurrence, or the stomach seems to be upset. The
quantity can be gradually increased until the kid is having
a pint per feed four times a day at three months old.

From birth to ten days or a fortnight the kid should
be fed five times a day, the hours between each feed being
even, except the last feed at night, which should be as
late as possible. At a fortnight one feed can be dropped,
but a little more milk given at the other feeds to make
up. Good times for the first fortnight are 8 and 11 a.m.,
2, 5, and 9 p.m. After this, 8 a.m., 12 noon, 4 and 9
p.m. are best.

Regularity is important, and once your times are fixed they should be adhered to. To allow a kid to become too hungry, when it will take the bottle too fast, is to invite tummy trouble.

Kids will nibble grass, greens, hay, and earth (which is very good for them) from an early age. They usually begin to eat concentrated food at about a month old. A little bran, flaked maize, crushed oats, and the dust of linseed cake should be put in a shallow vessel. They will soon learn to eat it. As soon as they begin to eat this solid food they will chew the cud.

A bunch of sweet hay should be put in a low rack or tied up in a bundle and hung at a convenient height, when the kids reach a month old. Make sure they cannot get their heads or legs tied up in the string. Branches of nut, elm, ash, etc., can also be hung up for them. A very small quantity of cabbage, kale, etc., may be given, but only when quite dry and not touched by frost.

An iodised mineral salt lick should be available from the start. You will notice the kids going to their " salt cellar " and having a good lick after a bottle.

Apart from bottle-rearing, kids can, of course, also be taught to feed from a bowl or bucket, thus saving time and expense with bottles and teats. A drawback with this method is that they may drink too quickly, become " pot-bellied " and scour. To induce the kid to drink, let it suckle a finger dipped in the milk, and gradually draw its mouth down to the milk. A very few lessons like this and the kid will drink of its own accord.

Kids must never be coddled, and must have plenty of exercise. In spring and summer they can stay out in a small wired enclosure containing a box or barrel into which they can creep to sleep. Or you can let them have the run of a shed or an ordinary chicken house, for shelter and sleep, a large hay-lined box being inside for sleeping. Any house used for kids must be draught-proof and leak-proof, but be airy and well-lighted, and facing south, sunshine being essential to their growth.

For exercise, a strong box turned upside-down in their pen for them to jump on makes an ideal " castle," and when they are a few weeks old a bridge made from two upturned boxes and a plank rigidly fixed gives these delightful little animals much pleasure.

A word about milk substitutes. These can be used

when there is a scarcity of natural milk, or if you have a kid to rear, with no nanny available. There are several well-known makes on the market, and directions are supplied for their use. The cost works out roughly at 3d. per gallon. Normally, the substitute should be gradually introduced to supplement the milk, after the age of one month.

### REARING A MALE KID.

If you have a well-bred male kid to rear, for the first three months of his life he can be kept with the female kids and treated exactly as they are.

Do not stint him with milk, but at the same time don't overdo it. Four pints of milk a day are sufficient for any kid, and three pints will be found to be enough for most kids. Too much milk weakens the digestion and the kid is apt to be " pot-bellied " and the flesh soft and flabby.

Encourage him to eat concentrates and hay as early in life as possible, give him as much concentrates as he will eagerly clear up (within reason, for some kids *will* eat more than is good for them, but I find as much as they eagerly clear up a good general rule) and as much hay as he can eat.

Exercise is most important for good growth but anyway while he is with the other kids he will get plenty if they are provided with boxes or a fixed barrel or something similar on which to play " king of the castle."

Teach him to lead well on a halter while he is young, to know his name and to come when he is called. These apparently minor points will save a lot of trouble when he is older.

It is a mistake to play with any male; jumping up and so on may be amusing when he is small, but remember that it will be you, and not he, who will know the difference when he weighs about one and a-half hundredweights !

The male kid will probably be used from six months onwards, and so during his first " season " he wants perhaps even a little extra care so as to keep him growing well. As the season approaches, his appetite will fall off badly; try and tempt him with a change of rations, or add a little treacle to his concentrates. Try bread crusts, flaked maize on its own, a little spice in his food, or anything you think he will eat. If, however, he definitely will eat but very little, don't worry unduly, for they soon pick up again once the season is over. The great thing is to have your male in really first-class condition at the beginning of the season. See that a mineral lick is always available.

Don't treat a male as a complete outcast. If there are no goats in season he can quite well accompany the females on their walks with you, if he is equipped with a halter. Try to amuse him a little while he is alone in his yard, for boredom is the chief cause, I believe, of objectionable habits. Give him leaves to pick over, branches to bark, and a section of an old tree trunk—or something as strong —will be welcome to jump on.

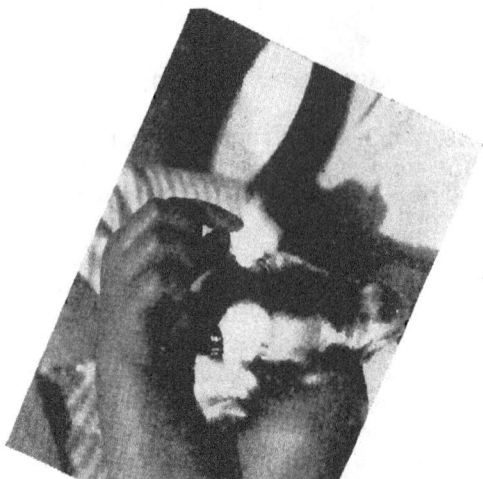

horns can be prevented by a simple "operation." Instructions are given on page 35, and should be studied in conjunction with these photographs.

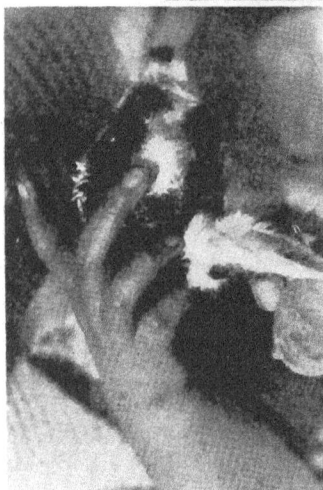

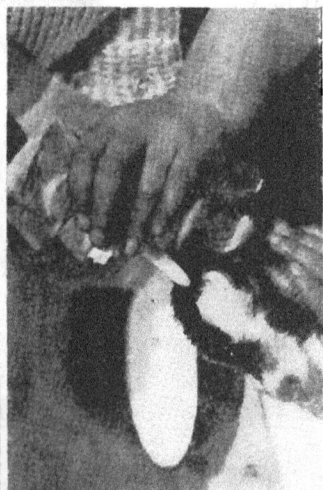

DISBUDDING

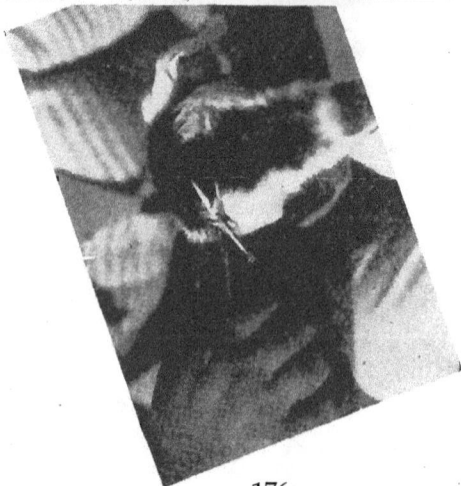

The family goat should preferably be hornless, for safety's sake. The majority of goats are now bred without horns, but if a kid is going to be horned, the growth of the

176

## Disbudding Kids.

WHETHER a goat has horns or not is immaterial from a milk production point of view. A horned goat is likely to be as good a milker as a hornless one. But it is a matter of practical advantage to have goats without horns, especially family goats. Accidents can happen even in play with the quietest animals.

The horns of goats can be prevented from growing by disbudding kids at a few days old. This operation can be performed by anyone who is prepared to take a due amount of care. It should be done at not later than three or four days old.

Your kids may not necessarily be going to grow horns. One or more of them might be, however, and you should be prepared.

A method of telling if a kid will grow horns is to wet the animal's head and smooth back the hair from above the eyes to the top of the head. A hornless kid's hair will remain smooth; otherwise, the "curls" which will be present will reappear. Head shape is also a guide. A horned kid will have a flat forehead, not rounded. Another method to determine if disbudding is necessary is to try to move the skin back and forth over the forehead. If quite freely movable no horns are developing; if the skin seems grown fast to a prominence horns can be suspected.

In time, of course, the "buds" can be felt developing, and later, little horns.

To disbud, all hair should be clipped away from the horn bud, a tiny patch of clear skin showing a pin-head point, and vaseline smeared around it, on the forehead and ear bases, to prevent any possibility of the disbudding agent running and burning. Special proprietary disbudding "sticks" are sold for the purpose, and should be used with care, the fingers being protected with brown paper.

The kid held firmly between the knees, the stick should be damped on moistened blotting paper and carefully dabbed on the bud and immediately around it in a half-penny-size circle. Repeat the dabbing for about a minute, occasionally moistening the stick.

Having treated both buds thus, rest the kid, then treat each bud again until it is rather black, blistery, and soft. As the bud becomes softer, use greater care not to break the skin. Give a final light dabbing after another interval, and finish by gently pressing a small wad of cotton-wool on each place to absorb surplus moisture. Afterwards place the kid by itself and see that it does not rub its head, when it will quickly forget the operation. Make sure also that the mother cannot get at the kid to lick its head.

# THE KIDS

**Breeding.** Wild goats *kid* in the spring, when there is fresh green food to make milk and to feed the young things as they grow up. Domesticated goats are descended from wild ancestors, and the majority of kids are born in March and April, though a few may be as early as January or as late as July. Very occasionally a goat will kid in the latter half of the year.

A young animal, such as a kid, is born as the result of a mating between its male and female parents several months previously. The length of time that elaspes between mating and birth is called the *period of gestation*. With goats it is, on the average, 150 days.

During these five months the kid is developing, inside its mother, from a tiny *egg-cell* into a completed animal. The materials needed to make this growth are provided by the blood of the mother, and it is not until the kid is born that it starts an independent existence, breathing with its own lungs, taking in food with its own mouth and digesting it in its own stomach.

An in-kid goat, with her living burden inside her, must be treated gently, although plenty of exercise and fresh air are good for her. She has her young to nourish as well as herself, so she must be well fed and kept in good, hard but not fat condition.

The egg-cell from which a new individual develops is one of many produced by a pair of organs (called *ovaries*) inside the body of the female. Another name for an egg-cell is an *ovum* (plural : *ova*). These ova are not available for development at all times, but only at intervals of three weeks during the mating season (August to February). For a few days at each of these times several ova are produced, and the goat is then said to be *on heat*, or *in season*. The outward signs of heat are a general restlessness, continuous bleating for no apparent reason, frequent wagging of the tail, and sometimes, if the goat is in milk, a temporary reduction in the yield. These signs are not equally obvious in all goats.

An ovum cannot develop into a kid until it has been *fertilized* by a male reproductive cell (*sperm*). This fertilization occurs at the time of mating, and the goat will refuse to

179

mate unless she is on heat. As soon as the goat has been successfully mated (i.e. one or more ova have been fertilized) she will cease to come into season. She should, however, be watched for signs of heat three weeks later. If they appear, she is said to have *turned* or *returned*, and must be mated again. If they do not she may be assumed to be in kid.

A goat in season should not be left out of doors entirely without supervision, for a male may, by instinct, be drawn to her from a distance, and an unwanted mating thus take place.

If no male goat is kept on the farm a journey is necessary when a goat is to be mated. If the journey is made on foot the goat must not be hurried, and she must be allowed to have a rest of half an hour or so before making the return journey. If she is taken by car she must be carefully handled, and she should be rugged up to prevent a chill.

Two or three months after the successful mating, the growth of the *embryo* (developing ovum) will show itself in the shape of the goat's body. Still nearer to the time of *parturition* (giving birth to the young) the udder will begin to be enlarged. The goat is then said to be *bagging up*. Shortly before kidding there may be a considerable quantity of milk in the udder, though under ordinary circumstances this should not be drawn off.

The diet of an in-kid goat should be *laxative* : that is to say, the food should pass easily through the digestive system, and the droppings should be softer rather than harder than the normal. Green foods, roots, bran mashes and treacle are laxative foods.

The best place for a goat to kid is a loose-box that is familiar to her. She should certainly not be tied up. If the box has to be a strange one, accustom her to it a week or two before she is due to kid. The box should be clean, and it is a good practice to disinfect it before it is used for a kidding goat.

Most people leave the nanny alone while her kid is being born, as in the majority of cases she will manage the event quite easily. It is as well, however, to keep an eye on her so that we can be assured that conditions are normal. If they are, the kid will be born so that its two forefeet appear first, and with its nose resting on the forelegs. If, after birth, it immediately shows signs of life, and if the mother soon starts to lick it, we can be sure that all is well.

A kind of bag protects and supports the kid during its development inside the body of its mother. At birth this bag breaks, and the broken remnants pass out of the goat's

20. Goats, especially the young ones, like to clamber about, and a rough platform, or a few boxes and a plank, give them opportunities for play and exercise. The triangle of sticks, attached to each kid's collar, prevents the kid pushing through the large mesh wire fence that surrounds the paddock. These triangles are called either puzzles or pokes. They are more often used for adult goats, and if made for kids must be very light in weight.

21. A barrel makes a useful shelter for kids on free range. A board nailed across the front prevents the barrel rolling and retains the straw inside. If a newly born kid is taken from its mother and is kept at night in a large loose-box, it should be provided with a box or barrel for greater warmth. How old do you think these kids are ? What is their breed ?

body a short while later, and usually within an hour of the birth. These remnants form the *after-birth* or *cleansing*. It is important that the goat should get rid of this. Most stockmen' watch for it and remove and burn it, otherwise the nanny may eat it. Although this is quite natural it is generally considered dangerous for a domesticated goat. If there is more than one kid at a birth the appearance of the after-birth is a sign that the last kid has already arrived. Twins are as common as single kids, triplets are not unusual, and four or five at a birth occasionally occur.

The unborn kid receives food and oxygen* from its mother, and these are conveyed along the *navel* cord that joins the blood streams of the two animals. When the kid is born this cord breaks naturally and there is a small amount of bleeding. The cord soon heals, however, and in a day or so shrivels up and falls off, leaving a small mark, the *navel*. Some stockmen dress the navel cord, as soon as the kid is born, with carbolized oil, or iodine, as a precaution against infection and to hasten the drying of the cord.

A goat is always thirsty after kidding, and she should be given a drink of *chilled* (with the chill off) water. When she has washed her kid she can have a bran mash, and, later on, a little hay and some fresh green food. Her feeding should be light for the first three or four days after kidding, the quantities being increased and the usual concentrates added gradually until a normal milking ration is reached.

When the mother has washed her kid, the youngster, warmed and invigorated by the process, will usually attempt to get to its feet in order to suckle. The struggle to do so is good for it, and, though a little help may be given, the young thing should be allowed to make most of the effort for itself. At first the little legs buckle under the weight of the body, and sometimes the feet are turned back from the pasterns, but after a little practice and a drink or two, the normal healthy kid "finds its feet".

Some kids need help in finding the teats, especially if these hang very low. Some goats do not take to their kid at first, and it is difficult to persuade them to stand while the kid suckles. Patience is needed for handling such a case, but patience nearly always wins. In very difficult cases the kid must be bottle-fed from birth. Indeed, many goat keepers do this always. See next page.

The milk given by a goat for the first 2 or 3 days after kidding is different in composition from normal milk. This

---

* The gas in the atmosphere that is used for breathing.

first milk is called *colostrum*, or *beastings*, and the kid must have this if it is to thrive. If there is more of it than is needed for the kid it can be used for making custard, for it is quite fit for human consumption.

**Rearing.** There are several different ways of rearing kids, and each way has its advantages and disadvantages.

*Suckling.* The kids take their milk direct from the nanny from the time they are born until they are *weaned* (cease being fed on milk). If the milk is needed for the house or for sale the kids must be weaned early : say, at 2 or 3 months old. They will probably, however, grow into better animals if they are allowed milk until they are six months old or more. Weaning, whenever it comes, must be done gradually, the kids being allowed less and less milk, and this at longer intervals, as weaning time approaches. They must also be encouraged to take more dry food to compensate for the smaller amount of milk. With care and skill they may be weaned without suffering any check to their growth.

The disadvantages of rearing kids on the goat are that it is not easy to ration the kids, it is impossible to know exactly how much milk the goat is giving, and it is difficult to sell a suckling kid separate from its mother.

The advantage, however, lies in the saving of labour, for the kids help themselves to the milk.

*Bottle-feeding.* This is sometimes called *hand-rearing*.

Some people allow the kid to take the colostrum direct from its mother, and start bottle-feeding on about the fourth day. Others feed the kid on the bottle from the beginning and do not allow it to suckle at all. The younger the kid the easier is it taught to drink from a bottle.

A baby's feeding bottle can be used at first. The valve can be removed from the one end and the other end fitted with a teat such as is used by shepherds for their lambs. Teats that have been used are softer, and therefore better, than new ones. Heat the bottle in hot water, then wrap it up in a piece of flannel and fill it with the freshly drawn milk. With practice, the correct temperature is arrived at without further heating or cooling of the bottle. Now back the kid into a corner and gently insert the teat into its mouth. In most cases it will quickly learn to suck, though some kids are slower at learning than others. It must not be allowed to get the milk too quickly, yet the milk must come without excessively vigorous sucking. The flow can be regulated by means of one's thumb placed over the upper end of the bottle. When the kid is about a fortnight old the baby's bottle can be changed for an ordinary, narrow necked wine bottle fitted with a soft rubber calf teat.

If the kid is to be bottle-fed from birth, the goat can be allowed to clean and dry it with her motherly tongue, but as soon as this is completed the kid is removed out of reach, but not out of sight. The goat's udder is then *eased* (a little milk is drawn) and the kid receives its first drink. Side by side, mother and child will probably be quite contented and will not fret. For the first few days milk should be drawn from the goat whenever the kid needs feeding. After that, the normal twice daily milkings can start and the milk be warmed for the kid's feeds. After 2 or 3 weeks of bottle-feeding the kid usually forgets the possibilities of its mother's udder, and the two can be safely left together. To make doubly certain the kid may first mix with the goatlings, who will strongly resent any attempt by the kid to suckle. The kid soon learns to keep off.

For bottle-feeding the following average quantities of milk are needed :

| Age of Kid. | Number of Feeds per day. | Approx. quantity at each feed. |
|---|---|---|
| Up to 1 week | 4 or 5 | $\frac{1}{4}$ to $\frac{1}{2}$ pint. |
| 1 to 3 weeks | 4 | $\frac{1}{2}$ to $\frac{3}{4}$ pint. |
| 3 to 6 weeks | 4 | $\frac{3}{4}$ to 1 pint. |
| 6 to 12 weeks | 3 | 1 to $1\frac{1}{2}$ pints. |

After 8 weeks the milk can be gradually reduced and its place taken by skim milk (plus cod liver oil) or by some good brand of calf meal (gruel). The milk must not, however, be entirely replaced by substitutes until the kid is about 3 months old. From then onwards for the next 2 or 3 months, the quantity of liquid given can be gradually reduced, and not more than one "bottle" a day need be given after 6 months. This reduction is continued until the kid is feeding on dry food only. The exact quantities of milk, or milk substitute, must depend upon the kid's appetite, but no kid should receive more than 5 pints a day or more than $1\frac{1}{2}$ pints at one feed.

A mixture of equal parts of linseed cake, bran and flaked maize, together with a little of the best hay, can be fed from the age of a few weeks, the quantities being very small at first and gradually increased. A kid in the company of older goats learns to eat solid food and to browse more quickly than if it is kept alone or with others of its own age.

This method of hand-rearing is by no means the only one,

22. Kids must have their bottle feeds at strictly regular times. The temperature of the milk should be 90 to 100° F. Too high a temperature is probably worse than one too low. The bottle must be kept perfectly clean, for otherwise the milk will turn sour and upset the kid's digestion. To what breed does the kid in this picture belong? Note the ear carriage.

23. A kid should be given the opportunity to learn to drink water when it is quite young. Half the weight of an animal's body is due to the water it contains, and water is constantly being given off in the breath, perspiration and urine. In winter the goat will drink more if the water is chilled, i.e. has the chill taken off. Note the chain-link fence.

nor would it suit all circumstances or be approved of by all goat keepers. Some say that a very much better kid can be reared if more whole milk is fed for a longer period and at more frequent intervals. Sometimes whole milk is fed to a kid until the end of its first winter : that is, until it is about a year old. When concentrates are short it may be best to ·feed the kids liberally on milk in this way, though the maxi-mum figures, given above, must not be exceeded.

Between these two extremes (weaning to gruel at an early age and weaning direct to dry food at a later age) there is room for many variations of method, and on many of them a strong, healthy goatling can be reared. The truth of the matter is that a good stockman can rear an animal success-fully in one of many ways.

If we are starting goat keeping by purchasing a couple of kids, and if these kids are, when we buy them, being bottle-fed on goats' milk, some substitute must be found for this milk since we ourselves have none. The change over, however, must not be sudden, and this needs the co-operation of the person from whom we are buying the kids.

**Treatment of the Nanny after Kidding.** She must be fed lightly for a few days after kidding, and not be milked right out until about the end of the third day. Otherwise, she may develop *milk fever*. See page 46. A freshly kidded goat should not be allowed out of doors unless the weather is dry and mild.

Goats are generally mated, and therefore kid, once·a year. The effect upon a milking goat of being once more in kid is to encourage her milk to dry up, and if she is not mated in the first season after kidding she may continue to milk for another year. By treating one or other of our goats in this way it may be easier to maintain a winter supply of milk. It is not unusual to find goats that have continued in milk for two or even three years because they have not been mated.

A goat should be dry a few weeks before kidding again, and this can be encouraged by not stripping (see page 37). Some goats are easier to dry off than others who may continue to give a little milk right up to the time of kidding.

**Disposal of the Kids.** The female kids can be reared to become, in their turn, mothers and milkers. If their dams, grand-dams and other female relations are good milkers we say that the kids *have milk in their pedigrees*. If, in addition, they are strong, well-built and of dairy type (see Pictures Nos. 13 to 16), they should, if possible, be kept in the herd. Most goat keepers are unwilling to sell their best female kids, for they need them to take the place of the milking goats as these grow old. (The average length of life of a milking goat is 10 to 12 years.) We know most about the goats that we ourselves breed and rear, and this gives home-reared goats a special value.

A female kid born early in the year (January) is sometimes mated at the end of the same year. This is her first breeding season and she is still only a kid (less than 12 months old). Breeding at so early an age is only wise if the animal is very well grown and has *good bone* (see page 17), for otherwise its growth may be permanently *retarded* (kept back). The alternative, which is the much more usual plan, is to delay mating her until the next breeding season. This is about a year later, and the animal will be a goatling about 18 to 22 months old. The late-born (April-June) kids are nearly always mated in the following year.

The female kids that we do not wish to keep can be sold, or they can be killed for meat at 6 to 10 weeks old. The male kids that are not wanted for breeding can also be killed for the table at this age. The weight of a newly born kid varies within wide limits, but the average is about 8 lb. Singles weigh more than doubles, triplets or " quads ". If well fed, the live weight of a strong single kid at 8 weeks old is about 25 lb. This will provide about 14 lb. of meat.

Fewer males than females are needed, for one adult male is enough for mating with 50 females. A great many surplus males are, therefore, born, and only the very best (see page 19) should be kept for breeding. Some people destroy surplus billy kids as soon as they are born, and if this leaves the mother without kids she must be milked ": little and often " for the first two or three days. This imitates the action of the kids.

Male and female kids ought not to be kept together after they are 3 months old, otherwise they may breed, and this at so young an age will stunt their growth.

**Feeding the Goatlings.** Hay and green food, with roots in season, can form the basis of their diet. If concentrates are scarce they must be reserved for the milking goats, but young, growing goatlings benefit by 1 to $1\frac{1}{2}$ lb. per day of a ration balanced for milk (see page 23). They should be kept in good condition but not made fat. Goatlings and kids repay time and trouble spent upon their upbringing, for their capacity to give milk in later life depends very much upon their early treatment.

**Feeding the Billy.** A male goat under one year is a

male *kid* ; between one and two years old he is a *buckling* ; and when he is two years old he is called a *billy* or *buck*.

All breeding animals must be kept in *hard* condition. This means that they must be in perfect health and, though well covered with flesh, lean rather than fat. A balanced diet, exercise and fresh air are, therefore, as necessary to the male as to the female goats. Hay, green food and roots can be fed as to the other goats, and, up to the age of 2 years, a ration of concentrates suitable for the milkers. At 2 years old most goats are fully grown, and a billy can then be given a con-centrate ration less rich in protein and more rich in starch than before. When concentrates are scarce he will have to receive less of these, though he must not be allowed to get down in condition, *especially just before the mating season starts in August.* During the mating season most male goats have a very poor appetite, and this generally causes them to lose weight. Under careful management, however, they soon regain condition once the mating season is over.

All animals must be treated gently if they are to do well and if their tempers are not to be spoilt. An adult male goat needs firmness in handling, for however good his temper he is a powerful creature and can be, at times, a handful.

24. Anglo-Nubian male. A male goat, of any breed, should be well grown and well developed for his age. He should have a wide, deep chest, well-sprung ribs, and strong hind-quarters. "Quality" is essential if the male is to transmit milking qualities to his daughters. Quality is difficult to define in words, but one of the indications of it is a soft and supple skin carrying short hair of a soft texture.

# Kidding

THE normal gestation period of a goat—i.e. the period during which she carries her young—is 150 days. Sometimes the kids are dropped a few days early, sometimes a few days late.

Some weeks before the kids are due, if the doe has been kept in a tie stall, it is well to remove her to her kidding pen so that she may move about freely. Have the pen clean and well bedded. Several weeks before the kids are born she will show a noticeable depression at either side of her tail and a hollowness at the hip bones. Her udder will fill, gradually at first, then toward the end rather rapidly. Feel the udder from time to time and should it become hard and shiny it may be necessary to milk out a small quantity of milk. So long as the udder is flexible, no milk need be removed.

A day or two before kidding the doe will become restless, lying down and getting up frequently, arranging her bedding and talking to you. There may be a mucous discharge for a day or two which just before kidding will become heavy and gelatinous. Have your equipment ready and convenient—newspaper and towels for drying off the kids, boric acid powder, sterilized scissors and a deep box or carton well bedded with hay or straw or even strips of newspaper in which to place the kids; a basin, bar of soap, a soft wash cloth and soft towel for the mother. Daytime kidding is a great relief, but even if it means sacrificing your sleep, plan to be on hand at the time of birth. It will give

"Capricious" comes from *capra* originally a fantastic goat leap—but only those who have watched kids at play grasp the full meaning of this word.

Saanen Kid just born.

Saanen Kid just 10 minutes old!

the doe confidence if she is nervous, and you will be ready to give assistance if necessary. Nine times out of ten no help will be needed.

In a normal birth the front hoofs of the kid will show first with the head between. When the kid's head emerges fully break the sac in which it is enclosed so that it can breathe and wipe the mucous from its nose and mouth. You may help the mother, if she needs help, by pressing on her sides as her body contracts, and you may grasp the head of the kid and help to ease it out with the mother's efforts, but do not pull the kid from her. Such haste may result in hemmorhage.

Usually the cord attaching the kid to the mother breaks of itself, but it may be necessary to use your small scissors. Snip it several inches from the kid's stomach and tie with a soft cord two or three inches from the navel. Be careful not to tie the cord so tightly that the tissue is cut. Sometimes it is advisable to place a tight sterile binder around the kid's body.

The second kid usually follows in a few minutes, although I have had a doe whose second kid was born eight hours after the first. Sometimes there are three kids, or even four.

After a half hour, if the mother seems comfortable and free from pain it is fairly safe to assume that there will be no more kids. She may then be given a dish of bran mixed with quite warm water, moist but not wet, a drink of warm water and a supply of fresh hay. Then the kid can be attended to—dried off so that her hair is soft and fluffy, the navel dusted with boric acid powder and the kid placed back in her box. If the barn is cold pin old blankets around the stall to protect the mother and kid.

Sometimes the kid will come hind legs first—a little more difficult birth for the mother, but not alarming. And again a kid may be crosswise in the uterus and need to be turned before it can be born. If you do this turning yourself, be sure that your nails are clean and your hands sterilized or that you use a rubber glove, and be very gentle. Also it may be, more rarely, that there

Quadruplets!

is a dead kid which the mother cannot expel. In this event it is wise to have the services of a veterinarian unless you are confident of your own ability to cope with the matter.

After the kids are born make the mother comfortable by removing the soiled bedding and putting down a clean layer. But do not disturb her to the point of a complete cleaning of her pen at this time. Wash her udder with warm, soapy water and dry it with the soft towel. She is then ready to be milked or to feed her kids.

The waste matter, or afterbirth, may come from the mother a half hour or so after kidding. If it does, thank Lady Luck and go about other duties with a feeling of relief, for the kids can wait some hours if necessary for their first feeding. It more often happens, however, that it is several hours before the afterbirth is expelled, and as some does are inclined to eat the afterbirth it is well to check on the mother from time to time so that you may dispose of it when it is free. But here again, do not pull it from the mother. Give nature a chance to do the job unassisted. If the mother eats the afterbirth the most serious effect seems to be a lack of interest in her food for a day or two which can often be corrected with a warm drink to which bicarbonate of soda is added.

One of the astonishing things about new born kids is the fact that often as soon as they are dried off, about a minute after birth, they stand up and walk. Unlike a number of other animals—puppies and rabbits, for instance, kids come into the world fully equipped—eyes wide open, tiny teeth, and muscles well coordinated. And there isn't any animal more appealing than a frisky, two-days-old kid.

## Ailments.

OF all domestic animals the goat is perhaps the least liable to disease, and some of the worst of those that attack others leave the goat unaffected. It is, of course, liable to certain ailments, and is more susceptible to trouble in certain circumstances, such as inter-breeding, coddling, mis-managing in regard to rearing, feeding, and so on. But ordinarily, a constitutionally sound goat properly managed will give its owner few worries. It is said that goats in Britain are becoming less hardy, but this I contend is a question of upbringing. Many caprine ailments can well be treated at home, and the goat-keeper should deal with minor illness himself when he can. It is no disgrace, however, but a duty, to seek skilled veterinary attention—blessed phrase! —when required.

Some of the commoner complaints are here dealt with: —

*Accidents.*—These will happen, even in the best regulated goat herd. Common sense is the best prevention. For instance, do not provide hay in nets for kids; they may become strangled by the netting. Do not tether a goat near a retaining wall with a drop the other side. 'Ware barbed wire. Keep horned and hornless goats apart.

Every goat owner must keep a small medicine chest handy, and know how to use the contents to best advantage. A knowledge of simple first-aid can be acquired readily these days !

*Blood-in-the-Milk.*—When milk is found to be streaked with blood, this may be the result of a chill, blow, or strain causing rupture of some of the small blood vessels in the udder, or, according to one authority, too rich feeding causing over-stimulation of the mammary glands. Gentle milking, a dose of Epsom salts (1 tablespoonful to a ¼ pint of water), the temporary cutting down of the rich ingredients of the ration, a bran mash every other day, and liberal green food will speedily cure. I have frequently found that this condition indicates the need for toning up the whole system by giving a course of iron tonic.

*Chills and Colds.*—A drink of half a pint of warm beer, or a quinine tablet, a sack or rug over the body, and a spell in a draught-proof stable will nip many a chill in the bud. The symptoms develop in the obvious way—a staring coat, shivering, hunching of the back, and later the usual runny nose and lost appetite. Proprietary medicines are available for treatment in advanced stages.

*Constipation and Indigestion.*—A good milking goat is a big eater and is therefore more prone to digestive disorders, and should have treatment at once before it becomes serious. If her appetite is not quite so keen, or she has "lumpy" droppings, something is wrong. I find the following treatment mostly puts matters right:

Get one pint of pure linseed oil from the chemist and give her two tablespoonfuls of this daily for, say, a week, or until she is normal again. If "blown" through gorging on wet greenfood, a teaspoonful of pure turpentine can be added to the oil. During treatment give plenty of good hay and water.

*Diarrhoea or Scour.*—This distressing trouble is usually a symptom of some other disorder, such as chills, worms, wrong feeding, etc., and the cause must, of course, be searched for in order to ensure permanent cure. Kids may scour because of overfeeding, or receiving milk at too high or too low a temperature, or lack of cleanliness. There are several remedies: Arrowroot biscuit ; a teaspoonful of prepared chalk in milk ; white of egg in a little water; dry flour baked in a pan until browned, mixed to a thin paste in warm milk with a little starch added to the first dose, and fed to the kid a little at a time at frequent intervals.

With older animals, all that may be necessary to remove the irritant matter is a dose of castor oil, varying according to age from a teaspoonful to four ounces, following with a little prepared chalk (up to a tablespoonful for an adult) or arrowroot biscuit. If the diarrhoea persists, give from a small teaspoonful to a dessertspoonful of the following mixture every eight hours:

Compound tincture of morphia and chloroform, 4

drachms ; liquid bismuth, 4 drachms ; oil of cloves, 1 drachm ; cooled linseed tea, 7 ounces.

Linseed tea is made by putting ¼lb. of whole linseed in a saucepan with half gallon of water and boiling very slowly on side of stove all day, then straining.

Veterinary chemists, of course, keep prescriptions to treat diarrhoea that they can make up for your goat.

If diarrhoea is due to worms, treat for worms as described on page 50.

*Foot Trouble.*—If your goat goes lame it may be because the hooves want trimming, or a thorn or stone is causing discomfort. But if the feet are hot and tender it may well point to your goat having too rich a ration and too little green food. An old-fashioned remedy for lameness of this type is a mixture of lard and sugar. Bran and linseed meal poultices are another form of treatment. At the same time, the diet must be cooling, and a dose of linseed oil given daily.

*Flies.*—When these are a worry to goats, one of the many proprietary fly dressings may be applied along the back. Oil of citronella, oil of lavender, and a wash made by macerating walnut leaves in vinegar, are also useful.

*Garget, Caked Bag,* or *Mammitis,* as it is variously called, is the principal udder trouble you must guard against, and what it really amounts to is inflammation of the udder. The causes are many, but chiefly ill-health, or lying about on cold, damp ground or wet grass, or injury to the udder.

The first sign is the hardening of the udder, which may become quite lumpy, and curdly milk drawn. A proprietary " udder drench " should be given the goat and she should be placed in a warm, dry, well-bedded loose-box. Hot fomentations should be applied to the udder, which should be particularly carefully dried afterwards so that there is no risk run of a further chill. The udder should also be well massaged, and all the milk or curd that can be extracted should be. The udder should then be well rubbed with one of the recognised udder salves, or an ointment made of camphor mixed with lard. This treatment should be repeated every two or three hours until the udder returns to normal.

Treated promptly in the early stages, the goat will soon be as right as rain again, but if neglected you stand the risk of losing your goat.

E

*Lice.*—These are not likely to trouble the family goat kept well-groomed and cleanly, but should they appear, tobacco powder, or one of the proprietary lice powders, should be dusted into the coat the wrong way of the hair.

*Sore Teats.*—Hard, cracked or chapped teats will be quickly soothed and cured by the application of udder salve. A tin of one of the many preparations available should always be kept handy. You need not fear that they will taint the milk. A soothing lotion which can be made up at home, if preferred, consists of boracic acid, 10 per cent. in solution ; a similar quantity of glycerine, and water. Put in a wide-mouthed jar and dip the teats into it after each milking.

*Tainted Milk.*—The most frequent causes of unpleasant flavour in milk are that the goat is in poor condition, is suffering from indigestion or worms, " strong " foods such as turnip or cabbage have been fed just before milking, or insufficient care has been taken over cleaning milk utensils. These suggest their own remedies. In the latter case one is particularly liable to get " almond-tasting " milk, and I cannot over-emphasise the importance of thoroughly scrubbing all utensils with *cold* water first before they are scalded out with hot water. Even boiling water will not cleanse milky things unless cold water is used first. As an experienced goat-keeper once said to me : " I found, to my surprise, that washing up dairy utensils was a real ' job ' to learn."

*Worms.*—A goat with worms will develop poor condition, her breath will be offensive, and her coat staring. There may be recurring diarrhoea, and worms may actually be seen. There are several excellent goat worm cures on the market, and you will find it quite simple to dose your goat and get rid of the parasites. If a goat is wormed regularly in spring and autumn there will be little danger of the parasites gaining a hold unsuspectingly and pulling the animal down in condition. The following is my own method :

I use a well-known make of worm pills, giving the goat two at a time (comprising one dose), late in the evening. I give a light feed of hay only about 8 p.m., following with the pills at 10 p.m., or later if possible, and giving no further food till morning. Sometimes the pills can be hidden in stale bread, and the goat takes them well. But the best method is to straddle the animal (as though riding horseback), raise her head up and back with your left hand, force jaws open gently with your right hand, push pill well down the throat, and then, still holding head up, close mouth and nostrils with right hand. The goat is then compelled to swallow. Remember to keep the head up and insert each pill separately. This dose can be repeated in about a month's time. Say this is done in the early autumn, it should be unnecessary to worm again until the spring, or six weeks after kidding. Never worm an in-kid goat.

The worms may be expelled from 12 to 36 hours following dosing. If the goat is stall-fed or in her stable at the time, be sure to remove *all* droppings and burn them.

After worming, a course of tonic will put the goat in fine fettle again. For tonics, see page 54.

# DISEASES.

OME of the older breeders supposed that the Angora was not subject to any disease, but as goats have been introduced into new territory, they have become affected by some of the same troubles which bother sheep, but usually to a less degree. Some of the worst sheep diseases, such as scab, do not bother goats, but the goat has some special complaints which do not effect sheep. Very few carcasses are condemned by the government meat inspectors at the large packing centers. Tuberculosis is almost unknown.

## LICE.

Nearly all goats are infested with lice, a small reddish louse, a goat louse. Lice rarely kills the animal infested, but they do annoy the goat greatly. Goats will not fatten readily, and the mohair is usually dead (lusterless), if the animals are badly infested. It is an easy matter to discover the lice. The goats scratch their bodies with their horns and make the fleece appear a little ragged. On separating the mohair the lice can easily be seen with the naked eye. The best means of ridding the goats of this annoyance is with almost any of the sheep dips.

A dip which does not stain the mohair should be selected. The goats should be dipped after shearing, as it does not take much dip then to penetrate to the skin. One dipping will usually kill the lice, but the albuminous coat covering the nits (eggs of the louse), are not easily penetrated, and it is usually necessary to dip again within ten days, so that the nits, which have hatched since the first dipping, will not have a chance to mature and deposit more eggs. Goats can be dipped at almost any time, but if in full fleece they will require a larger quantity of liquid, and if the weather is very cold, there is some danger.

### STOMACH WORMS.

Stomach worms affect goats, and in some instances their ravages prove fatal. There are a variety of these worms, but the general effect on the animal is about the same. They are usually worse in wet years. The goats affected become thin and weak. They usually scour. Sometimes the worm, or part of the worm, can be found in the feces. These same symptoms are caused by starvation, so the two should not be confounded. There are many drenches in use for the treatment of this trouble, and some of the proprietary remedies have given some relief. Goats running on dry, high land are rarely affected.

Verminous pneumonia of sheep may also occur in goats.

## FOOT ROT.

Foot rot is a disease which effects both goats and sheep, if they are kept on low wet land. It rarely proves fatal, and can be cured if the cause is removed, but it sometimes causes a good deal of trouble. The goats' feet swell between the toes and become so sore that the animals are compelled to walk on their knees. It can be cured by carefully trimming the feet and using solutions of blue stone. Goats should not be put on wet land.

Sometimes the glands of the neck enlarge, a condition known as goitre. This is sometimes fatal with kids, but usually cures itself. There is no known remedy for it, but it is comparatively rare.

Anthrax, tuberculosis, pleuro-pneumonia and menengitis, will effect goats, but these diseases are very rare. Some of the southern goats have swollen ears, but what the cause of this trouble is no one has yet determined.

## POISONS.

There are several plants which will poison goats, but very little is known about them. Some of the laurel family are responsible for the death of a good many goats yearly, and some milk-weeds will kill if taken in sufficient amount at certain times of the year. These plants should be avoided as much as possible. Treatment has been rather unsatisfactory. If the poisoned animal is treated at once, an active purgative may rid the system of the irritant. Epsom salts and crotin oil have given relief.

Mr. Schreiner describes an epidemic of pleuro-pneumonia which destroyed many flocks of Angora goats in South Africa. The disease was effectually stamped out in that country, and it has never appeared in American flocks. Mr. Thompson has described a disease called Takosis, which was supposed to have caused the death of many goats in the Eastern States, and along the Missouri River Valley. Some claimed that this trouble was caused by change of climate, others thought that it was starvation or lack of proper care. There is very little evidence of it now in the United States. All in all, the Angora goat is the healthiest of domestic animals.

# HEALTH

**Prevention of Ill-Health.** The goat is a hardy animal, and good health is the rule in herds that are well managed. By good management is meant :

Breeding from vigorous animals that are not too closely related to one another.

The use of clean, sheltered land for the goats.

Good feeding.

Dry, well-ventilated houses and good beds.

Cleanliness in all things.

Many accidents can be prevented by a little forethought. It is dangerous to leave a halter or headstall on a goat on free range without supervision ; to tether on a steep bank unless the pin is at the foot of the slope ; to tie up a goat in her stall with too long a chain ; to leave forks and other sharp tools lying about ; to cut the ties of a truss of straw or hay and leave the string where somebody can eat it ; to leave the doors of store-rooms and lids of food bins open so that the goats are able to gorge themselves ; to leave gates open or insecurely fastened so that the goats trespass and find harmful food. These are dangers that are always present but easily avoided.

**Symptoms of Ill-Health.** A sick goat loses some or all of the signs of health. See page 8. If the animal is feverish its temperature will rise above the normal (102-103° F.) and the base of the ears and the feet will be cold. (A goat's temperature is taken in the rectum.) The state of the droppings is a reliable indication of the state of the animal's digestion. If they are soft and green, and the food passes through too quickly, the animal is said to *scour*. If there is a partial or complete stoppage the animal is *constipated*. Normally, the droppings should be firm and dark brown in colour. Indigestion may cause *loss of cud* (the goat stops chewing). When the milking goat is ill or sickening, her yield of milk is reduced. When an animal loses its bloom, and becomes tucked up and thin, it is said to *lose condition*.

These symptoms may be produced by some slight disturbance inside or they may be the early indications of more serious trouble. When, by experience, we become quick to notice that a goat is a little off colour, we shall often be able, by simple treatment, to prevent serious trouble developing. By simple treatment is meant such things as :

(a) making the diet more laxative* ;
(b) reducing the concentrates and making the ration as digestible as possible ;
(c) giving chilled water in place of cold ;
(d) keeping the suspected invalid away from wet grazing ;
(e) increasing the comfort of her stall ;
(f) rugging her up ;
(g) paying particular attention to cleanliness.

If the patient does not very soon respond to this simple treatment, consult the veterinary surgeon.

**Nursing.** The recovery of an invalid depends a great deal upon the nursing she receives. The basis of good nursing is strict attention to the ordinary rules of *hygiene* (health), an understanding of the veterinary surgeon's instructions, and a sympathy with the patient. The more we learn the better nurses shall we be, and the best teachers are experienced stockmen and our own daily work amongst goats. Although many facts and ideas and the explanation of technical terms can be found in books, such learning must not tempt us to over-estimate our own cleverness. It is right to have confidence, but there is always a great deal more to learn.

**Possible Troubles.** The following are the most common :
*Worms.* All domestic animals can harbour parasitic worms, and these are most commonly found in the stomach and intestines. It is, however, very easy to worry too much about these, for a few are, perhaps, no bad thing. When, however, they increase in number to such an extent that they cause the goat to lose condition, the matter becomes serious. Before this happens something must be done to get rid of them. Even when passed in the droppings these worms are too small to be seen with the naked eye, but their presence is shown by the goat's poor condition.

If goats are kept under crowded conditions, grazing the same small piece of pasture year after year, the soil may become heavily infected with worms. The eggs of the worms, which pass out in the goat's dung, can live in the soil for a long time, and thus they tend to accumulate in a small paddock. The land is then said to be *goat sick*. It is possible that grassland becomes goat sick more easily than heath and woodland.

* A bran mash is laxative. To make this, scald half a pound of bran with enough boiling water to moisten it thoroughly. Cool it with cold water until it is like warm, sloppy porridge. It is improved by the addition of a teaspoonful of salt.

6. The easiest way to measure milk is to weigh it. A gallon of goats' milk weighs, pproximately, 10 lb. 4 oz. The empty pail should bring the pointer on the scale to the ero mark. Each goat's yield should be entered on a record sheet, either daily or once week. If a goat yields 115 gallons in a year, how many pounds of milk has she given?

7. Drenching means giving liquid medicine. Stand the goat in a corner. Use a narrow, ong necked bottle. If the goat coughs, remove the bottle from its mouth and lower its ead *IMMEDIATELY*, for if the medicine gets into the windpipe serious trouble may be aused. Whereabouts in a goat's mouth are the incisor teeth? Is it the same in all animals?

If, however, the goats are not crowded on their pasture, or if they get frequent changes of land ; if the houses are disinfected once or twice a year ; and if we do not buy trouble in the form of a worm-infected goat, the danger is not great. Treatment consists of drenching the goats with a medicine that clears out the worms.

The worms that infect sheep can also infect goats, so pastures that have been closely grazed by sheep for many years are not suitable for goats.

*Scour.* This affects bottle-fed kids more often than the other goats. The most common causes are too much milk, or milk taken too quickly or at the wrong temperature (see page 33), or uncleanliness of the bottle or teat. The simple treatment is to miss one or two feeds and then to give reduced quantities of milk until scouring ceases. The white of an egg in a tablespoonful of water can replace one of these feeds.

It is not unusual for all goats to scour slightly when the spring grass comes. See page 22.

*Constipation.* Make the diet more laxative by including more green food and replacing one of the feeds of concentrates by a bran mash. Treacle is laxative, and a little of this may be added to the mash. If the constipation persists give a dose of castor oil in linseed oil. The dose for an adult goat of average size is 4 ounces of each oil, mixed well together.

*Chills.* These are upsets caused by the animal getting chilled. The symptoms vary, but the most common ones are a general listlessness, a staring coat and a loss of appetite. The goat should be comfortably boxed out of the draught, and rugged up. Do not, however, exclude fresh air and make the goat house hot, for this will hinder recovery.

A safe and often used drench (see Picture No. 27) is half a pint of warm beer. If this treatment does not quickly cure the patient, and if her temperature rises, consult the vet.

*Poisons.* Very few plants are invariably poisonous, but numerous ones occasionally cause trouble. This makes diagnosis difficult, and the vet. should be fetched when poisoning is suspected.

Rhododendron poisoning causes *vomiting* (sickness), and pieces of the leaves may be brought up and recognized. Immediate treatment is necessary if the goat's life is to be saved, so, while waiting for the vet. to arrive, give a dose of one large teaspoonful of bicarbonate of soda in a tablespoonful of melted lard. This dose can be repeated half an hour later. Skill is needed to keep the lard melted without making it too hot for the goat to take.

28 and 29. The horn, or wall of the hoof, is trimmed level with the sole of the foot. If the hooves become overgrown the goat will have difficulty in walking and will be unable to graze in comfort. Tender feet always lead to loss of condition. An assistant should hold the goat during this operation. There is no need to seat a goat on its hindquarters in the way that a shepherd handles his sheep.

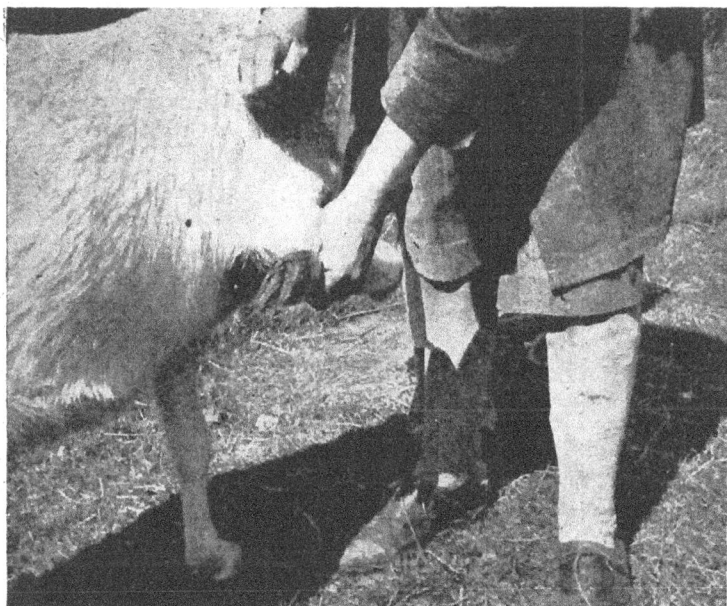

*Milk Fever.* When a goat comes into milk there is a sudden removal from her body of all the materials that form the milk. These include quite a considerable amount of *calcium*, and the removal of this causes the symptoms known as milk fever. They will appear not immediately, but some months after kidding. The goat shows signs of extreme exhaustion, and is unable to walk or even, perhaps, stand. Her treatment is a matter for the veterinary surgeon, and until he arrives all that can be done is to keep the goat warm and prevent her lying flat on her side. Bags, loosely packed with straw, can be used to prop her up in the normal lying down position.

The treatment usually consists of an injection of calcium, made with a hypodermic syringe. Rapid recovery often follows, but careful nursing is needed for some days afterwards.

*Skin Parasites.* If goats are kept under clean conditions and do not get run down in health, they are very seldom troubled by these. Regular grooming and the fortnightly use of flowers of sulphur will usually keep the skin and coat clean.

*Wounds.* A bottle of mild disinfectant, some lint and a few bandages should be kept at hand in case of accidents.

**Care of the Feet.** The hooves grow, just as horses' hooves and human finger nails grow, and must be cut. Running about on hard ground helps to keep the feet in good condition, but cutting of the horn is needed too. If neglected the feet may become diseased and permanently injured. They need attention about once a month, although it is better not to treat a goat that is heavy in kid. Pictures Nos. 28 and 29 show how the goat is held, but the actual method of cutting the hoof can only be taught by demonstration and learnt by practice.

**Disbudding.** No kid, when first born, has horns, for these, if they develop at all, do so during the first few weeks of the kid's life. It is not always easy to tell, at birth, whether the kid is going to grow horns or not, though if it is going to be horned there is usually a suggestion of curls over the places where the horn buds will be. Sometimes these buds can be felt quite easily, and at 3 days old there is usually no doubt about the matter.

If we do not want the horns to develop, the buds can be treated with caustic potash or some proprietary brand of material. This is generally done when the kid is 3 to 5 days old. The operation is described fully in several of the books mentioned on page 47. It needs very great care, and is not recommended by everybody.

# EMERGENCIES

While the services of a competent Veterinarian are not in the least discredited and quite to the contrary are urgently suggested when needed, it is also advisable to be prepared for emergencies, such as sudden illness, or accident or the absence of your Doctor of Veterinary Surgeon, thus possibly saving a life or at least preventing prolonged suffering of the patient. With that thought in mind, the following simple remedies are recommended and it is suggested as a precautionary measure, such as may be kept on hand, which do not deteriorate with age, also such instruments as pill gun, 2 oz. Syringe and others of common use may prove of value equal to many times their cost.

# REMEDIES

A few prescriptions from a reliable practitioner are here noted for your convenience and safety.

### Cough and Cold Mixture

Chlorate of Potash .......... 4 drams
Chloride of Ammonia ....... 2 ounces
Tincture Iron Chloride ...... 4 drams
Fluid extract Stramonium .... 1 ounce
Glycerin ................... 1 ounce
Water to make ............ 1 pint

Mix. Shake before using. **Give adult goats 1 tea-**spoonful three times daily.

### Digestive Tonic and to promote appetite

Powdered Nux Vomica ...... 1 ounce
Powdered Gentian .......... 1 ounce
Bicarbonate of Soda ........ 4 ounces

Mix thoroughly. Give adult goats 1 teaspoonful twice daily.

## For Worming Goats

Copper Sulphate (Blue Stone)
Crystals ................. 1 ounce
"Black Leaf 40" ........... 1 ounce
Boiling Water ............. 3 quarts

Dissolve the Blue Stone Crystals in the 3 quarts of water. Allow the solution to cool. Strain. Then add the Black Leaf 40. Give goats from one to three ounces of the above solution, depending on size and age of animal. Give with metal dose syringe after fasting overnight. Dose the animal slowly to avoid strangling. The dose should be repeated in three or four weeks.

## For Eye Infections

Yellow Mercuric ........... Oxide 1%

The above is quite useful in clearing cloudiness of eyes which is due to simple injuries or infections from dust and weed pollen.

## For Indolent Ulcers or Sores

which do not heal, Allantoin Ointment 2% is quite beneficial applied once or twice daily.

## For Arthritis or Swollen Joints

Iodex Ointment applied daily and rubbed in thoroughly.

# REMEDIES

## For Ringworm

Paint the infected spots once daily with Tincture of Iodine for several days.

## For Diarrhoea

| | |
|---|---|
| Bismuth Subnitrate | 2 ounces |
| Powdered Catechu | 3 ounces |
| Sodium Bicarbonate | 4 ounces |
| Powdered Charcoal | 4 ounces |

Mix thoroughly. For kids give one teaspoonful and for adult goats one tablespoonful three times daily.

or

| | |
|---|---|
| Milk of Bismuth | 2 ounces |
| Paregoric | 2 ounces |

Shake together and give kids one teaspoonful and adult goats one tablespoonful three times a day.

or

| | |
|---|---|
| Tincture, Krameria | 2 ounces |
| Paregoric | 2 ounces |

Kids one teaspoonful three times a day.

## Oil Dressing for Wounds

| | |
|---|---|
| Carbolic Acid (Liquid) | 1 ounce |
| Oil of Tar | 4 ounces |

Turpentine ................ 2 ounces
Linseed Oil to make ........ 1 quart

Paint on wounds to promote healing and repel flies and maggots.

### For Goat Pox (Local Treatment)

Paint the vesicles and ulcerations with the following mixture after each milking.

Tincture Iodine ............ 1 ounce
Tincture Benzoin Compound . 1 ounce

### Ointment for Swollen Udders

Extract of Pokeroot ........ 10%
Camphor ................. 2%
Petrolatum ............... 88%

Apply morning and evening after bathing of swollen parts with warm water.

### For Chapped and Cracked Teats

Oil of Thuja .............. 6%
Zinc Oxide ............... 13%
Petrolatum ............... 81%

Apply after each milking.

# REMEDIES

## *Remedy for Scours*

4 oz. each of the following: Ground Ginger, Powdered Chalk, Ground Catachu.

Mix and feed one teaspoonful night and morning and omit grain until well.

## *Mineral Salt Mixture*

One pound each of common baking soda
Sulphur
Charcoal
Saltpeter
Air slacked lime
Crushed tobacco
$\frac{1}{4}$ lb. clear crystals of pulverized Blue Stone thoroughly mixed.

Then add above to 15 lbs. common barrel salt and again thoroughly mix and keep dry, serving same in a convenient corner of the stable accessible at all times when the goats will use it as needed.

# Keeping Goats Healthy

IF YOU have read much about goats you have probably met time and time again the statement that goats are the healthiest of domestic animals. This does not, however, justify their neglect nor the thought that they can be left entirely on their own and come out on top. Goats are healthy, but only if they are cared for and given good food. In damp and drafty quarters, or left out in storms or cold winds with no shelter they have every chance to develop pneumonia, and it is pretty hard to save the life of a goat with this disease. If by some unfortunate chance a goat gets chilled, a good brisk rubbing is in order, a woolen sweater or blanket, and a warm drink. If 'she won't take the water, add molasses or something sweet to make it inviting, or try warm goat milk or even black coffee, and give her a warm bran mash. If she doesn't improve promptly have the veterinarian without delay.

Another unwholesome condition to which goats are prone is worms, and these, too, may prove fatal. There is a large variety of worms to which both sheep and goats are susceptible and their cause is chiefly too concentrated pasturing. Consequently, when goats are wormed they should be given a fresh pasture or the pasture treated so that the larvae are destroyed. Your State Department of Agriculture will advise you as to the most up-to-date chemical to use as well as how long it will be before goats can be returned to a pasture so treated.

Noticeable indications of worms in a goat are loss of appetite, thinness, pale eye and mouth membranes, rough coat, and soft droppings. A specimen of the droppings may be sent to your own state experiment station or college of agriculture for analysis to confirm any suspicion you may have that your goats are wormy.

For some years the most effective treatment employed for worms in sheep and goats was a drench of copper sulphate solution, but more recently the approved treatment is the use of phenothiazine. The copper sulphate required the fasting of the animal and was dangerous to administer because it had to be given as a drench. The phenothiazine may be given as a powder in the food, requires no preliminary fasting; some goats will eat it with their grain unsuspectingly, but some will not touch the feed. If liquid phenothiazine or phenite is used it may be administered with a bulb syringe or the pellet form can be placed down the throat. If you can get your hands on what is known as a small "balling gun" the pellet will go down more easily.

The United States Department of Agriculture in an article "Phenothiazine for the Control of Parasites of Farm Animals" states that this drug is effective against such worms as common stomach worms, lesser stomach worms, bankrupt worms, hookworms, large-mouthed bowel worms, and also nodular worms for which previously no effective treatment had been found. The dosage for a mature goat is 20 grams ($\frac{1}{5}$ of an ounce). For animals weighing 60 pounds 12 grams is sufficient. As a preventive two treatments a year are considered adequate—in the fall and in the spring before the animals go out to pasture. The advisability of administering drugs to pregnant animals is always questionable, and recent experiments have shown that dosing does pregnant more than two months may cause them to abort. Dosing during the first two months of pregnancy is less risky. The one disadvantage in the use of phenothiazine is that it turns the milk pink for three or four days and unusable as human food. However, the treatment of the animals may be so arranged that but one milker at a time need be treated. If the dry goats

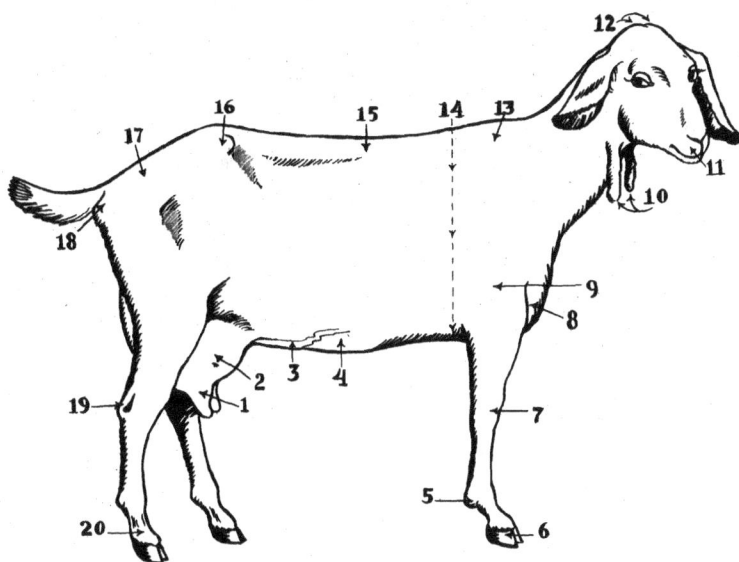

Diagram Showing Parts of a Goat's Body:

| | | | |
|---|---|---|---|
| 1. teats | 6. hoof | 11. muzzle | 16. hip bone |
| 2. udder | 7. knee | 12. knobs | 17. rump |
| 3. milk vein | 8. chest | 13. withers | 18. pin bone |
| 4. belly | 9. shoulder | 14. heart girth | 19. hock |
| 5. claw | 10. wattles | 15. back | 20. pastern |

and the milkers are pastured in separate enclosures, a trough may be placed for the dry stock in which a mixture of phenothiazine and granular, loose salt is available at all times—nine parts salt to one part powdered phenothiazine, by weight. Then the semi-yearly treatment may be omitted for the dry stock.

Experiments at the University of Missouri have shown that the phenothiazine-salt mixture is quite effective *in controlling* parasites of goats and sheep, provided the animals are comparatively free of the parasites at the time they are put on the mixture. The fall and spring treatment is adequate, provided the grounds are not heavily infested with parasites and the phenothiazine-salt mixture is used continuously.

Another serious condition that you may encounter with a milking doe is udder trouble. This may be caused by injury in the pasture. You should see that there are no objects in the pasture such as pieces of wire or stumps sticking out of the ground, or projections from edges of buildings, or any object in the lot where goats are kept that might injure the udder. Also the entrance to the barn and milking stand should be arranged so as to cause the goats no difficulty. Many injuries to the udder may thus be prevented.

Injury may also be caused by overcrowding the doe's udder with milk either through irregularity in milking, through too heavy grain feeding after freshening, or through failure to remove excess milk if her kids are nursing and not taking her output. Twice a day gentle massage of the udder and hot applications are advised to relieve the congestion, plus milking every two hours for sixteen continuous hours and application of a lubricating ointment. The doe must be kept warm and given plenty of clean, dry bedding, and the treatment repeated at frequent intervals until she has recovered. If the doe does not improve quickly under this treatment it is safest to consult the veterinarian as the condition may be due to infection and if neglected might result in the loss of her ability to produce milk, or even in death, and also might spread to other animals in the herd. When any trouble of this kind occurs it is advisable to discontinue the feeding of grain or other foods that stimulate milk production, to discard the milk from the animal, and to milk her last in the string of milkers.

Scours, or diarrhea, may be corrected usually with a dose of oil followed by boiled milk (half the usual amount in the case of a kid being fed milk) to which, if the case is stubborn, browned flour may be added.

In general, goats respond to the same sort of medication for minor digestive disturbances as is given to human beings—baking soda for indigestion; a soap and water enema for constipation; Epsom salts to cleanse the digestive track of irritating substances which may cause scours. For indigestion or impactation, five heaping tablespoonfuls of Epsom salts in a quart of water are given to an adult goat. Repeat in five hours if no action is obtained and at the same time place a tub of fresh water before the animal and withhold all feed for twenty-four hours. If at any time a goat refuses her grain, this should be considered a danger signal that impactation is developing and immediate steps should be taken to prevent further serious symptoms. The next feeding of grain should be withheld and the hay reduced in amount. Very often this will avert an attack of impactation or indigestion.

Occasionally, goats, particularly if kept for a prolonged time in tie-stalls where the platforms become soaked with urine, will develop small eruptions on the teats and udder. These are stubborn things to clear up and can be spread from one to another by the milker. Washing with a mild disinfectant will help and the application of ointment such as is used for diaper rash in babies is effective.

Abscesses sometimes develop about the mouths or jaws of goats. They usually are not painful and are long in developing to the stage where they become soft enough to lance. They should be watched and when ready should be opened, the pus pressed out, and the wound washed with an antiseptic.

Lice sometimes occur, particularly in the early spring before the goat has shed her heavy winter coat. An application of lard and kerosene—two parts lard, melted, to one part kerosene—well emulsified—will usually dispose of these in the older goats. For the kids and pregnant does a non-poisonous powder, such as is used for dogs, is safer. Your feed dealer will have several brands from which you may choose. What appears to be the best treatment of all is a 5 per cent or 10 per cent powder of DDT rubbed into the hair; this is safe and effective.

Aside from unpredictable problems that may arise at kidding

time and which are not frequent, goats given reasonably good care and food will go through life with a clean bill of health—good insurance risks—and live to a ripe old age, barring accident. At present my herd includes a doe twelve years old, two does eleven and one ten. They are sound and healthy and active, have excellent teeth and none has ever been ill. They are still milkers, although giving less milk than in their prime. If you take proper care of your animals you have every right to expect as good results.

A surprisingly large number of goats die from accident, and in most cases these accidents can be avoided. A rope dangling from a hook where a kid can play with it is one way of inviting trouble; or a goat tethered with a thin, cotton rope or in a spot where she can tangle herself if frightened, another. If your goat is tied where you can't readily see her, visit her occasionally to be sure that she is safe. I've known of a goat eating a broom which caused her death, and the moral of that is: put your cleaning equipment in a closet or a room where the goats can't get at it, and keep it there with the door closed securely. Also put your grain some place where the goats can't gorge themselves on it when you are not around. That, too, can bring disaster. A water pail with the handle that swung over the goat's head as she drank and frightened her caused another fatality. It is much safer to use a galvanized tub or to have the handle of the pail removed. Bits of wire, or nails or glass get into the feed with tragic results, especially among kids—although this happens far oftener with cattle than with goats.

Get accustomed to anticipating what may happen and take steps to avoid it, and inspect your goats daily. Chances are that then you will detect any trouble before it becomes serious.

# Milk-O !

A BIG problem to the beginner is that of actually milking a goat, if no such thing has ever been attempted before. Here again you will find your worries disappearing as you get down to actual practice. It is an " art " readily acquired, and even children make good goat-milkers.

And what a satisfying business it is, this particular business of being your own dairyman. Off you go in the early morning. . . . There is an eager " Ba-aa! " in greeting as you approach nanny. Squish, squish . . . squish, squish . . . the warm milk frothing in the pail, gradually mounting—one, two, three, "Ah! three and a-half pints this morning, my dear !   I bet that's one up on old Smith.  Hope I meet him in the train!"

Some people squat on their heels to milk. You may find it more to your convenience to have a low bench on which the goat will quickly learn to stand, while you sit at the side on a stool. A suitable bench would be from 12-18ins. from the ground, the width 2ft., and the length approximately 3½-4ft.

A small bucket may be used for milking. Personally, I frequently use a 3-4 pint enamel mug, over the top of which I clip a thickness or two of butter-muslin, with a rubber band, as an additional aid to cleanliness.

The dairy business end of the goat is a delicate piece of machinery which must be treated as such. The udder must not be tugged or swung about or carelessly knocked. Treat it with firm gentleness.

The mechanics of milking are these: Grasp the further teat with the right hand, in the crutch between the thumb and forefinger. Squeeze, and with reasonable pressure follow up with the second and third fingers. If the teat is long enough it may be possible to use the little finger also. Relax the teat and begin again with the forefinger.

Perform the same operation on the other teat with the left hand, alternating with the right.

Note that there is no pulling or stretching of the teat, but only pressure. Occasionally with your wrist you should " bunt " upwards into the udder, in natural imitation of the bunting action of a kid when suckling. This induces a release of milk from the udder.

When you have extracted as much milk as possible in this way, you must " strip " the bag. This is most important. With each hand alternately, draw thumb and forefinger quickly up and down each teat, until the last drops of milk have been extracted. You need these " strippings " as they are the richest part of the milk. Gentle massage of the udder, particularly round the back, during this process, will be found to bring down more milk. Unless an udder is stripped right out at each milking the yield will gradually decline, so do not skip this part of the job.

Speed comes with practice, but always milk as quickly as possible. It ensures a steadier and increased flow, and avoids irritating the goat.

A goat should chew her cud contentedly during milking. If she does not, or is very fidgetty, endeavour to find out what is wrong. You may be pulling long hairs on or about the udder; or there may be a sore or chap on the teat.

For the first few days after kidding milk your goat particularly carefully and lightly; only take a little milk, but gradually increase the amount taken away, until you are milking " full out."

If your goat is a first-kidder she may give you a little trouble until she has become accustomed to an experience which is as strange to her as it is to you. You must be patient, petting and coaxing her, and treating the udder with care.

Very often the teats of a first-kidder are so small that they cannot be grasped easily with the hand. These can be gradually brought into better shape by gentle pulling and massage after applying lanoline or one of the proprietary udder salves. Persistent massage has worked wonders with many an unpromising udder.

### Making Your Own Butter and Cheese.

NOT least of the joys of keeping goats is that you can make you own butter and cheese. It is surprising how many people think that elaborate equipment is necessary for butter-making. This is far from being the case. Any goat-keeper with a fairly good supply of milk can make delicious butter with the aid of a few simple articles to be found in any kitchen.

I have known butter to be made regularly by being shaken up in a jam jar. I have myself made a weekly quota in an ordinary pudding basin, using a cheap geared cream-whisk. An average of 6 ozs. of butter from just over a gallon of milk has been obtained, with hand skimming. The milk, being skimmed, has still been quite creamy, and served its ordinary purpose in tea, etc.

The pudding basin method is this: After milking, the milk is strained, and cooled as quickly as possible (by standing in a bucket of cold water, if there is no other means), and set aside in shallow pie dishes for from 24 to 36 hours. A secret of obtaining good cream is to see that the dishes stand quite level, little wooden wedges being tucked under them if necessary.

The cream is skimmed off with an enamelled spoon, taking care not to bring away milk with the cream, and put aside in a bowl. Add a pinch of saltpetre and a *good* pinch of salt to every half-cupful of cream. Add salt with each subsequent batch of cream, stirring it in. In about four days sufficient cream will have been obtained to make the butter, and it will be at a good stage of "ripeness." Do not add fresh cream to your supply on the morning of making as this will not be sufficiently ripe.

The butter is made by whipping the cream, slowly at first and working up to a good speed. In a matter of minutes the cream will become granulated. A little cold water must then be poured in and the whipping continued until the butter grains are the size of wheat grains.

The buttermilk can now be poured off, to be used for scone making, and more cold water poured into the basin with the butter to " wash " it. Work the butter well with Scotch hands, flat pieces of wood, or wooden spoons previously stood to soak in water and rubbed with salt to prevent the butter sticking. Smack and work the butter until all the buttermilk is washed out, changing the water frequently till it finally comes clear. All that remains then is to beat the butter firmly to shape and extract all moisture possible.

To colour the butter is quite simple. During the whipping process two or three drops of butter-colouring annatto may be added. This can be purchased in small bottles from agricultural chemists. Or you may prefer to adopt the method now used by many goat-keepers of adding one or more egg yolks to the cream before churning. Not only does this give the butter a nice appetising colour but also makes it more nutritious.

Actually it is really no more difficult to make butter in a small table churn, which can be purchased from large stores and dairy supply firms. Naturally, more cream and better butter can be obtained with the aid of a cream separator, but these are a little too expensive for the family goat-keeper to consider, who will have nothing to be ashamed of in the quality of butter she can make in her pudding basin. Let me point out, too, that goats' milk butter has been officially declared an unrationed product, and may be sold freely without coupons.

## Goat's Milk Cheese.

There are one or two simple cheeses that the family goat-keeper can make, mainly of the cream or soft variety. Apart from these cheese-making requires rather more elaborate equipment than the average family can run to, and rather more attention to detail than could be found time for.

What I call " Family Cheese " is probably the simplest of all goats' milk cheeses to make. To one quart of fresh warm milk add half a teaspoonful of rennet, and stir well. Let stand for 12 hours, then cut the curd into pieces of uniform walnut size with a knife or enamelled ladle.

Transfer the cut curd to cheese cloth and hang up to drain for 24 hours. Scrape the sides to assist drainage.

Then take down, flavour with a little salt (it is a unique idea to use celery salt), working it in and pressing to shape by standing in a cup or bowl with a weight of some sort on top. Or use one of the proper little moulds which can be purchased from dairy supply shops. The mould should be stood on a small straw mat. Place another mat on top and slightly weight down. Keep thus for two days, turning the mould over and weighting from the other side once. The cheese can then be used.

Cottage Cheese is a well-proved recipe. This is made from skimmed milk, and if you have been making butter you will find it a profitable use for the milk left from the cream setting.

Bring the milk to a temperature of 75 deg. Fahr., and keep it there. Add " starter " or a little buttermilk to speed up the process of curdling, which should not take more than 24 hours. Cut the curd into 2in. cubes with an enamelled ladle and transfer it to a vessel submerged in another vessel of hot water. Raise the temperature to 100 deg. Fahr. " Cook " for 30 minutes, and stir at intervals of five minutes. When ready, place the curd on a cheese cloth to drain, and after ten minutes tie up in the cloth and hang it up to drain.

When the whey stops running out, take down the cheese and well work it with clean boards or Scotch hands, adding salt at the rate of 2½oz. to 10lb. of curd. It is not necessary to use rennet for this cheese, but some makers do so, and claim that a considerable flavour is imparted by it in time.

A simple cream cheese which can be made without rennet is this : Take some thick cream and cool it to 65 deg. Fahr., in three hours. Then hang it to drain in a draught but in a moderately warm place. Afterwards scrape down, and turn the sides to the middle at intervals of six hours. When nearly set add salt to taste, and mould up for use while still sweet.

For the following cream cheese, made with rennet, you require equal quantities of cream and new milk which should be mixed together thoroughly. Heat in another vessel of hot water, bringing the temperature to 80 deg. Fahr., then add 1 c.cm. of rennet to 1 gallon of mixture. After about an hour cut the curd into 1½in. squares, ladle it into cheese cloths, and drain as above.

A Bondon type cheese can be made, enabling you to use

up buttermilk, or sour milk. A mixture of two pints of either with 1½ gallons of whole milk will make about 12 cheeses in the regulation Bondon moulds.

Raise the temperature of the milk to 70 deg. Fahr., and add rennet at the rate of 1 c.cm. to 1 gallon, leaving to set. If the milk is set at night, the curd will be ready for ladling into a cloth for draining next morning. When the curd is drained change it into fresh cloth and tie up with a few pounds pressure placed on top for about 24 hours. Then mix in a little salt and place in the moulds, leaving the moulds to stand on straw mats until properly drained.

It may help you to know that about 17 drops of rennet equal 1 cubic centimetre (c.cm.).

### CORNISH SCALDED CREAM.

It will probably interest you to know how to make this, and the following excellent recipe was sent me by a lady in Cornwall: —

Put three or four inches of milk in an enamelled bowl, and leave about 12-24 hours, according to whether the maximum yield of cream is needed or the scalded milk wanted for use as soon as possible. Put the bowl over *gentle* heat. When a thick, wrinkled crust has formed and begun to crack, put the bowl in a cool place for about 24 hours. The cream can then be taken off the milk with a skimmer previously dipped in cold water. The scalded milk makes good puddings.

# MILK

**Milking.** 1. As a general rule goats are milked twice a day. The exceptions to this rule are :

(a) the freshly kidded goat, which must be partially milked out several times a day for the first two or three days if her kids are not suckling ; and

(b) the goat that is being dried off before kidding. When nearly dry she need be milked only once a day, and, later, every other day until dry.

2. The normal, twice daily milking must be carried out at fixed times, and the nearer these are to 12 hours apart the better.

3. There should always be peace and quiet in the goat house, especially at milking times.

4. The more quickly the udder is emptied the better. The action of milking must, however, be as even as possible.

5. Although there is a certain amount of vertical movement of the hands, the action that draws the milk from the udder is a *squeeze* and not a *pull*. The final stripping is done by drawing the fingers and thumb down the teat, but the less of this the better.

6. With the exceptions in paragraph 1, the udder must be completely emptied at every milking. If even a small amount of milk is left the goat is encouraged to give less at the next milking. Thus, incomplete stripping encourages a goat to go dry. Moreover, the last drawn milk is the most rich in butter-fat.

7. Some goats take to milking and *let their milk down* more readily than others. Much depends upon their handling as goatlings, yet in spite of the most gentle and wise handling some remain obstinate and hard milkers. The muscles at the top of the teat are controlled by the goat, and she has to be persuaded to relax these to let her milk down. With a freshly kidded goat the presence of her kid may help. A stubborn goat may sometimes be coaxed into letting the milk come by giving her a handful of something tasty at milking time, though if this practice is once started it will probably have to be continued.

8. If a goat that usually stands quietly suddenly becomes restive, her teats and udder should be examined for sores or

for long hairs that are being pulled. The hair on the udder should be kept short.

9. In the interests of cleanliness the milker should have dry hands. Some people say that milking with hands moistened with milk is better because it is " nearer to nature ". The fact that in many herds all the goats are milked successfully *dry-handed* shows that wet hands are not necessary.

10. Before milking starts, the hindquarters and udder should be cleaned. Any cloths used for washing and drying the udder should be frequently sterilized by boiling.

11. The first few squirts from each teat should not be drawn into the milking pail with the rest of the milk. This *fore-milk* is always dirty, for it washes out the *teat canal* (the passage in the teat down which the milk flows). It can be collected in a mug and given to the cats. If mixed with the bulk milk it will spoil the *keeping quality*. (See below.)

**Care of the Milk.** Milk turns sour by the action of *bacteria* (germs) that get into it after it has been drawn from the goat. If the goat is healthy, the milk (with the exception of the fore-milk) will be practically germ-free (*sterile*) when it leaves the udder, and the bacteria that get in afterwards accompany the tiny particles of dust that drop or blow into the milk. The effect of these bacteria is, sooner or later, to turn the milk sour.

The greater the number of bacteria present the sooner does the milk turn sour, and although a certain number do not matter it is most undesirable to have so many that the milk sours soon after it is drawn. The length of time that elapses before souring takes place is called the *keeping quality* of the milk.

Bacteria *multiply* (breed) more rapidly under warm than under cool conditions, so that if the milk is cooled immediately after it is drawn, and kept cool, it will remain sweet much longer. Small quantities of milk can be cooled by standing the container in cold water. All receptacles containing milk should be covered with muslin to keep out the dust.

The milk should be strained while still warm, for it then passes through the strainer more easily. The best type of strainer is one in which the milk passes through a disc of cotton wool. A fresh disc is used at each milking.

All utensils that are used for milk must be rinsed with *cold* water as soon as they are finished with. They are then washed with hot water to which a little soda may be added. Lastly, they should be sterilized, either by pouring boiling water over them or, in big dairies, by placing them in a steam chest. They should not be dried with a cloth but allowed to drain.

**Composition of the Milk.*** This is not the same with all goats, nor the same at both milkings, nor the same at all times of the year. 'An average composition is :

|  |  |  |  |  |  |  |  |  | % |
|---|---|---|---|---|---|---|---|---|---|
| Water | .. | .. | .. | .. | .. | .. | .. | .. | 87·73 |
| Fat (butter-fat) | | .. | .. | .. | .. | 4·50 | | | |
| Milk sugar (lactose) | .. | 4·08 | | | | | Total solids | 12·27 |
| Protein | .. | .. | .. | 2·90 | Solids-not-fat 7·77 | | | |
| Minerals (lime, potash, etc.) | 0·79 | | | | | | | |

                                                        100·00

The fat is present in the milk in the form of extremely small round particles (*globules*), and these, being lighter than the rest of the milk, rise to the top in the form of cream if the milk is allowed to stand undisturbed. The fat globules in goats' milk are much smaller than those in cows' milk, and this probably accounts for the facts that goats' milk is the more digestible and that its cream is slower to rise.

Goats' milk is whiter than cows' milk, but this does not mean that it is less nourishing.

There is a popular idea that goats' milk tastes and smells of goats. Provided that the goats are kept clean, this is not so. The male goat, however, is usually strong smelling (see page 6), so that when feeding and grooming him and cleaning out his shed it is a good plan to wear an overall kept for the occasion, and to remove this before returning to the goat house or dairy.

Cases of tuberculosis in goats are so rare that it is almost true to say that the goat is immune to it. Goats' milk has sometimes been accused of spreading the germs of Malta fever and the almost identical undulant fever. Under the conditions of goat keeping in this country, such infection need not be feared.

**Use of the Milk.** It is almost impossible to keep the yield of milk from a herd constant throughout the year. Because of the restricted breeding season milk is most plentiful in April and May and most scarce in January and February. The relative amounts taken for kid rearing, for use in the house, for butter and cheese making, and for sale, must be varied to suit the circumstances.

* In " Rabbit Keeping ", Booklet No. 3 in this series, the compositions of the milks of various animals are given. They can be compared with the analysis on this page.

The very small size of the fat globules makes butter making not easy, though for cheese making it is an advantage. Butter is made from the cream of the milk, and *skim milk* and *butter milk* are by-products that are very useful for feeding to stock. *Whey* is a by-product of cheese making, and, though not as nourishing as skim milk, can be fed to poultry and pigs.

The making of butter and cheese from goats' milk is dealt with in the leaflets mentioned on page 47 of this booklet.

25. Some people milk the goat as it stands on the ground, and others have a milking bench. This is about 4 feet long and 2 feet wide, and stands 1 foot to 2 feet above the ground. Goats soon learn to jump up on to the bench. This goat has a chain attached at one end to its headstall and at the other to a ring in the wall. This girl milks her goat from the left-hand side, but most people milk from the right.

# USE OF GOAT'S MILK

It has long been understood, even by those who had no experience, that goat's milk has certain

PROPER MILKING

positive advantages and certain special uses of the utmost importance. In chemical composition

goat's milk is not very different from cow's milk
as is shown by the following standard analysis:

|  | Cow | Goat |
|---|---|---|
| Fat ...................... | 3.90 | 3.80 |
| Milk sugar ................ | 4.90 | 4.50 |
| Proteins combined with calcium | 3.20 | 3.10 |
| Various salts (solids) ........ | 9.39 | .901 |

In vitamin content, however, goat milk is rather
remarkable. The following statement on this
point is taken from Dr. W. E. Krauss of the Ohio
Experiment Station:

Vitamin A—2000-3000 International Units per
quart
" B— 400- 500 Sherman Units per quart
" C— 30- 50 Sherman Units per quart
" D— 6- 12 International Units per
quart
" G— 35 Sherman Units per quart

However, goat milk has other very definite advantages. One of these is in the much smaller size
of the fat globules and it is thought that this physical fact has a great deal to do with the superior
digestibility of the product. It is known and gen-

erally recognized, for instance, that goat milk is particularly suitable for infants, invalids and convalescents, in fact for all who have any difficulty in handling ordinary milk and other customary

BOTTLING SYSTEM

foods. Moreover the goat milk is alkaline in reaction whereas cow milk is acid, and this may make a considerable difference with patients already suffering from hyperacidity. It is also rich in mineral salts, a fact which is specially important with young children.

Dr. Carl G. Wilson has said, "I am convinced that goat milk is the best substitute for human milk for infant feeding, not only because of its close similarity chemically and physically but also the readiness with which the infant's digestive organs receive and digest goat milk."

Another doctor has said, "When we seek a substitute for breast milk there is one, and only one, to offer, and that is goat milk. No matter what claims are made for advertised foods they cannot take the place of goat milk produced under sanitary conditions and fed properly."

A good many responsible medical men also believe that goat milk is particularly valuable in cases of tuberculosis where it is essential for the patient to receive and digest considerable quantities of good substantial food. It has been pointed out recently that Hippocrates, the father of medicine years ago, advised something like the modern treatment. With reference to a person afflicted with tuberculosis he said: "Let them live in high altitudes, drink goat milk and take sun baths."

In a recent publication, Dr. Charles E. Atkinson has said, "From a theoretical standpoint we see that goat milk has many points of superiority, but the much tried lessons of experience have proven that goat milk is actually a safer and better food. My work with patients has been altogether with tuberculosis, but in my close relationship with patients I have learned directly how goat milk has helped many babies. What has been said in regard to the infant also applies to the grownup person who is ill or who has a delicate stomach. I am convinced that if, in general, the physician's prescription for milk was made to read goat milk, digestive difficulties would be less common and the path to robust health made smoother."

Another physician, Dr. Harrison Crook of Washington, D. C., conducted an experiment in the use of goat milk for amoebic dysentery on which he reported as follows: "Referring to the case of amoebic dysentery in which goat milk was used, I have to say that the patient, after suffering a great deal for about three weeks, began to improve almost immediately when put upon goat milk. The nausea ceased almost instantly following its use. Previously the patient was weak, fretful, and moaned constantly, due doubtless to the lack of nourishment. The milk seemed to give strength and a very restful state was brought about. I cannot say that the dysentery was directly influenced by the use of the milk, but the patient's strength was preserved and she was thus enabled to more readily throw off the poison. I feel convinced that the use of goat milk will serve a good purpose in such cases when other food cannot be retained. Indeed I am so pleased with the results in this case that I shall not hesitate to use this milk in any case of general weakness."

The United States Department of Agriculture which has made extensive experiments in the breeding and milking of goats points out further that goat milk was found to have a much softer curd than the milk of cows. The curd tension of the Holstein's milk was about twice, and that of the Jersey milk about three times that of goat milk. This physical fact, also, has a very substantial influence on the digestibility of the milk.

It is further stated that goat milk is an ideal food for babies and convalescents and those suffering from anemia, nervousness, loss of weight, run-down condition, constipation, tuberculosis, ulcerated stomach, nervous indigestion, pernicious anemia, eczema, sugar diabetes, etc. Scientifically, goat milk is a therapeutic agent as well as a food.

Naturally the practical use of goat's milk in place of cow's milk appeals to the small householder, particularly the family of low income. Such a family can keep one or two goats and have sufficient wholesome milk for family use when it would be impossible for them to keep a cow. It seems likely that this simple, practical fact will have a great influence in the future in extending the use of goat's milk. Boiled right down to its essentials this is the reason why goats have been habitually kept very much more extensively in European countries than in America. In other words, the goat is the poor man's cow.

## Why Is Goat Milk Especially Desirable For Children

Because it contains the essential elements for proper development not only of general growth but especially is it found to contribute to the sound and enduring first and permanent set of teeth. This has been brought to our attention repeatedly by practitioners of dentistry, especially and frequently by general practitioners of medicine. This is easily accounted for by reason of the purity of goat milk from general infections, and this in turn is due to the superb health of the milk goat provided it has been kept properly as to sanitary quarters followed by strict sanitation in the handling of the milk, not only by the producer and distributor but the house maid who may per-

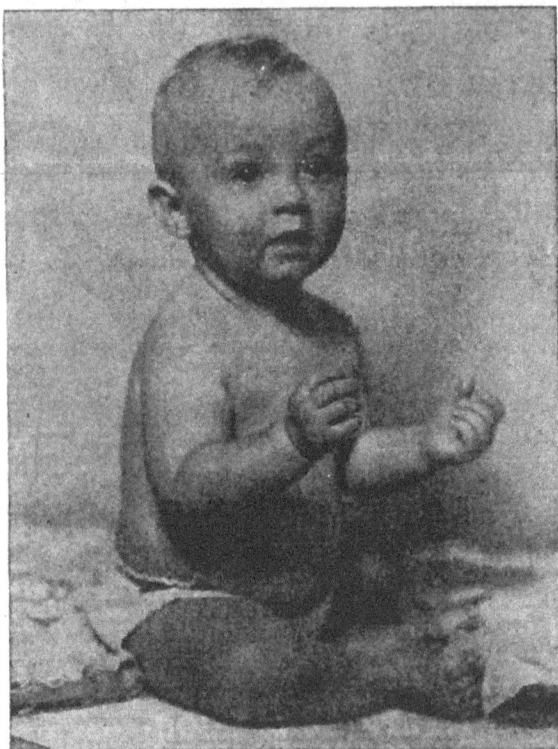

EVIDENCE OF SUPERB NOURISHMENT

mit it to stand on the door step for an indefinite period of time in the sun or even if no sun then during heated weather.

Also care should be exercised as to any milk being exposed to absorption of disagreeable odors in the refrigerator, such as onions and any and all foods which give off unpleasant odors as they impair the flavor of the milk leading the prejudiced to arrive at the conclusion it is because it is goat milk. I have on more than one occasion heard such criticism when I had purposely on the first delivery obtained cow milk for that one delivery easing my conscience by not charging for that one item.

It is an erroneous idea that goat milk must be treated by heat or otherwise. This is never necessary unless the child's stomach has become so inflamed that it will not retain any of the various baby foods which usually are tried. Those in charge, should at least dispel the sinful prejudice, and should accept the simple plain truth which applies to clean goat milk, as much as to pure spring water not requiring to be boiled before using.

It is a commendable fact that the up-to-date physician who is above retarding progress for the sake of long drawn out illness, with many calls and prescriptions, is more and more being recognized as worthy of confidence, while the opposite class which condemns goat milk is gradually dropping to the lower level where he regretfully belongs.

Briefly, the ailments most frequently benefited in infant feeding is where all other foods even including the breast milk of the nervous or ill mother disagrees, due to lack of digestion. Then and only then is it well to carefully avoid too sudden change, and by that I mean, do not expect that little stomach to accept an overdose but rather instead give half the amount previously given of other foods, gradually increasing as the case may warrant.

In adults the list of ailments which have responded to goat milk as the most easily digested, most nourishing and most desirable includes practically all cases generally classified as stomach troubles, which may be due to dissipation in eating as well as otherwise, resulting in ulcerations, abnormal growths, etc., which when relieved of such usually respond in time and frequently in a surprisingly short time. Advanced age usually includes one or more direct causes for ill health. It is urged not to classify goat milk as a medicine but nature's finest builder though even its good effects may be destroyed by indiscreet actions upon the part of the patient.

Exceptionally nervous people of any age find it advantageous. Those suffering from shock after accidents and those in a convalescent state likewise find it beneficial and those who lack strength from any cause usually bless its results.

There can be no solid reason for not using goat milk as freely as any other milk. In my lectures I have said I would not think of keeping house without goat milk any more than I would without a cookstove.

# OTHER USES OF GOAT MILK
## AND MILK GOATS

There are various ways in which goat milk can be and is being popularized such as butter, ice cream, candy, cheese both hard and soft, and new enterprises; along these lines are being brought into existence by those conveniently located to a generous supply.

Condensed and evaporated goat milk is obtainable, which has its advantages in cases of travelers going where fresh goat milk may not be easily secured. Also there are those who have not yet learned to bring their herds into normal year-round production, who find by keeping the milk in cans may be able to hold customers who otherwise would drift away.

Then there is another use for milk goats, (dead ones), for whether by accident or otherwise you occasionally will lose one and in that case if you will carefully skin the carcass and heavily salt the inside of the skin, wrapping it and shipping it promptly to a tannery you will find the return worth the time and expense for such makes most beautiful rugs and where treated for leather, it goes into gloves most satisfactorily, so all in all we do not claim quite as much as the packer does for the hog when he says he utilizes every part but the squeal, but we do claim there is a place for practically every part of the goat, under same circumstances.

# CHEESE, BUTTER, etc.

An almost endless list of names for different varieties of cheese may be obtained from the different countries which have down through all the generations since Adam, found goat milk and its by products in the form of cheese, butter, etc., well worth while. In many of the older countries there is a far more general use of same than in this country where vast acreage relieved the necessity of more densely populated countries, where every square front of productive land is utilized in some remunerative form, and the milk goat usually exists the major part of her time tethered along fences, highways, etc., except in such localities where mountainous tracts are pastured by herders. Under such conditions every drop is used either in its raw form or converted into cheese or butter and to answer the numberless questions asked the author, may it be said good clean goat milk (and that means produced under strictly sanitary conditions) properly and promptly cooled, may be converted into cheese and butter as aptly and satisfactorily as any other milk. For your convenience just one receipt for soft cheese (Cottage) and Hard cheese (Brick, etc.) is here given.

## Hard Cheese

Including every detail of sanitation and prompt cooling of the milk it is highly important to exercise the same scrupulous care as to the vessels and handling throughout all the way to the purchaser. The average temperature is approximately 86° F. The amount of rennet depends upon the amount of milk to which it is added, the object being to curdle the milk in order to permit the whey to be drained off, that the remaining solids may be made into cheese in the customary way. There are commercial colorings which may be used, which are harmless and inexpensive should you desire to give it a different shade for market.

## Soft or Cottage Cheese

This form is popular on the farm and requires very little suggestion, other than to say the average farm wife uses a clean cheese cloth bag permitting the prepared sour milk (avoid it being too sour) and if conditions are such as to require rennet which usually is the case, then do not hesitate to use it. The whey is disposed of by drainage through the sack which is suspended where the drippings may be conserved for the poultry yard or pig sty.

### Cream Cheese

This variety is made from good rich cream either naturally sour or soured by a limited amount of rennet, placed in a sack similar to that used in making cottage cheese. To these could be added many other forms of cheese but suffice it to say any person who appreciates the food value of good cheese will not be disappointed in being served with either form.

### Butter

Complaint is often made by those who fail to realize the making of butter successfully is largely due to the temperature of not only the milk but of the atmosphere as well, for unless general conditions are favorable, it is really exasperating to churn for an indefinite length of time only to be disappointed, whereas had precaution been observed in the beginning, all probably would have been quite different and highly satisfactory.

A good general rule, if you are inexperienced, is to consult some good reliable farm wife who markets her butter or go to the creamery where they convert the cream of the surrounding country into butter, ice cream, or to a candy factory, and ascertain the proper precaution to observe and as a rule they will regard it as good business to aid you.

Your butter made from goat milk will be made more attractive by adding a little vegetable coloring which is usually made from the juice of the carrots and that, as you know, is commonly accredited with having beauty ingredients when used in the human family.

# There's Milk in Your Backyard

MISS EVA LEGALLIENNE, a neighbor of mine, once told me of a guest she had for luncheon on a hot summer day who asked for a glass of milk.

"Goat milk?" queried the maid.

"Oh, no!" the woman answered. "I couldn't drink goat milk —even if Miss LeGallienne does."

Miss LeGallienne and the maid exchanged understanding glances and soon a glass of cool milk was set before the guest. A person accustomed to goat milk would have recognized its whiteness, but the inexperienced guest drank the milk and remarked, "My, that was delicious!"

"We never told her," said Miss LeGallienne, "and to this day she doesn't know she enjoyed a glass of goat milk."

I have had a number of such experiences. A woman wanted milk for her husband who had stomach ulcers. She was quite sure that he wouldn't drink it if he knew that he was getting goat milk, so she changed the cap on the bottle.

She worried when her husband tasted the milk and immediately demanded, "Where did you get *this* milk?"

"Oh, from a nearby dairy," his wife answered. "Why?"

"It's the best milk I ever tasted . . ." her husband replied.

It was a long time before she dared tell him he was drinking goat milk. But when she did he was so sold on it that he had no prejudice left.

I guess everybody who has kept goats has had some sort of similar experience. Ed Robinson, author of "The 'Have-More' Plan," tells about the time he was asked to give a luncheon talk before the Bridgeport Lion's Club. Even though it was in the main dining-room of the swank Stratfield Hotel, he brought along one of his goats, milked her, and had over fifty people compare goat and cow milk. About one-third said the goat milk was cow milk, another third said the cow milk was goat milk, and, of course, the rest guessed correctly.

Properly handled goat milk is almost impossible to distinguish from cow milk by tasting. Most of the prejudice against goat milk seems to come from people who have had it abroad where a picturesque herdsman milked it into a bowl. Perhaps his hands and the goat's udder were clean, but more likely they weren't. You know enough about milk and butter to know that even in your own refrigerator they absorb odors quickly. Also, some people in this country who have goats make the mistake of letting the male run with their milking goats. A buck, at certain times, *does* smell—and this odor is often picked up by the milkers. Moreover, goat milk, like cow milk, *must* be chilled immediately after milking—but more about this later on.

I have kept goats for many years—about fifteen to be exact. Obviously, I like them. But in this book I want to write about them objectively. I want to talk about all their good points— and their bad points too. For goatkeeping, like most everything else I know of, has its disadvantages.

First, let us consider the happy side of goatkeeping.

Milk and milk products account for 25 per cent of the average family's food budget. A goat is an extremely efficient small milk producer. A good goat should average two to three quarts of milk per day for ten months; the latter two months she should be rested before she has her young and, of course, begins another ten months' cycle of milk production. If two goats are kept you can have one milking at all times. Even if you buy all the grain a goat eats it costs 10 cents or less a day

251

This little girl finds Saanen kids excellent playmates.

to feed a milking goat; the sale of her annual kid or kids should pay for incidental expenses such as veterinarian fees, breeding fees, etc.

Goat milk is good milk. Today, some 60 per cent of the milk consumed throughout the world is goat milk. It is easier than cow milk to digest; it is naturally homogenized. That is, the fat globules, which are smaller than those in cow milk, stay in suspension. Many people, particularly infants, who cannot digest cow milk thrive on goat milk. A survey of people owning goats would probably show that a large percentage purchased their first doe because of some illness or allergy in the family, for from the days of Hippocrates, the father of medicine, down to the present, goat milk has been advised by physicians in the treatment of many human ailments.

A goat does not need as large nor as expensive quarters as a cow. Many people, particularly women who feel that a cow is too much to handle, can and do keep goats. Goats thrive on woodsy pasture; brush, weeds and poison ivy are manna from heaven.

Many families believe that they can use only two to four quarts of milk a day. Two goats should keep them supplied, whereas a cow that gives ten to twenty quarts a day would swamp them.

A goat is only a sixth the size of a cow and therefore can be transported to a veterinarian, or to a buck in the family car. Goats average from 166 to 202 births per hundred, depending on the breed; that is, your goat is more apt to have two kids than one . . . three kids are not uncommon and occasionally even four. A goat has a somewhat longer productive life than a cow. Goat milk sold retail brings a high price—25 to 60 cents a quart. It takes less capital to buy a couple of goats than a cow. Goat meat, called chevon, is good to eat and hard to distinguish from lamb.

Goats are friendly, intelligent, and responsive to human affection. The young goats are capricious, fun to watch, and

An idyllic street scene at Saanen in the Bernese Oberland, Switzerland. The goats shown in this picture are, of course, of the noted Saanen breed.

make excellent pets for children.

Now for some disadvantages. If you are going to keep goats, or any animal that has to be milked, somebody must be at home to milk twice a day. Milking should be done regularly. Goats are perhaps the most difficult of all livestock to keep fenced; they will jump or climb any fence less than 48 inches high. If they are not properly fenced they will get out and you can count on them browsing on your choicest young apple trees or shrubbery. It is as much trouble to make TB and Bang's tests on a single goat as it is on a cow. Because goats can be bred

The little boy who lives next door makes friends with some of my Nubian kids.

usually only in the fall and winter months it is more difficult to plan steady, year-round milk production. Even though goat milk brings more than cow milk, customers are more scattered and more expensive to locate. Generally, goat keepers do not keep records of production and it is harder to buy a good goat than it is to buy a good cow.

How many people believe the disadvantages are outweighed by the advantages? For the first time, the census in 1940 covered milk goats—and found 118,896 were "milked during any part of 1939 on 33,232 farms." The census goes on to say that "since 876,596 goats were enumerated that were not classified as Angoras (which are kept for mohair) and only 118,896 were reported milked in 1939, it is apparent that there is still a large population of goats classified as 'brush goats.' Their chief utility seems to be the clearing up of brush pastures, wood lots, and rough land, but they also contribute to the supply of kid and goat meat in southern and southwestern states. The number of goats milked covered only 3.6 per farm reporting. However, in some areas there were producing flocks of considerable size. Some of these larger flocks were adjacent to large city markets, but the largest ones were in the Southwest where much of the milk was used for manufacture of cheese."

Since over 100,000 people in this country see good reason for keeping milk goats—there must be something in it.

# Milking and Care of Milk
# and Equipment

MOST milking goats are milked twice a day, at intervals as nearly twelve hours apart as can be conveniently arranged. For such a task, repeated day after day, the right sort of equipment is essential. Otherwise it becomes irksome and what might be a pleasure becomes instead an irritating chore. With an expenditure of about ten dollars you can secure the equipment needed for milking and you would be wise to have it and to do the milking in a professional way. Your milk then will be as clean and wholesome as any dairy can produce, your disposition unruffled and time saved.

Instead of using that enamel kettle with only a small chip in it—which by the way can generate poisonous elements if used as a food container—get a seamless milking pail, preferably one with a hood to keep out particles of dirt and hairs; a milk strainer, and, for a larger dairy, a milk carrying pail (also seamless, which means no hiding place for bacteria). With the strainer you need a box of filter discs, made of sterilized cotton. One of these discs is clamped into the strainer for each milking —a fresh one each time.

It would be well also to have a strip cup. The first stream or two from the goat is milked into this cup which has at the top a very fine sieve. If the milk passing through the sieve leaves behind any lumps it usually means some unhealthy condition in

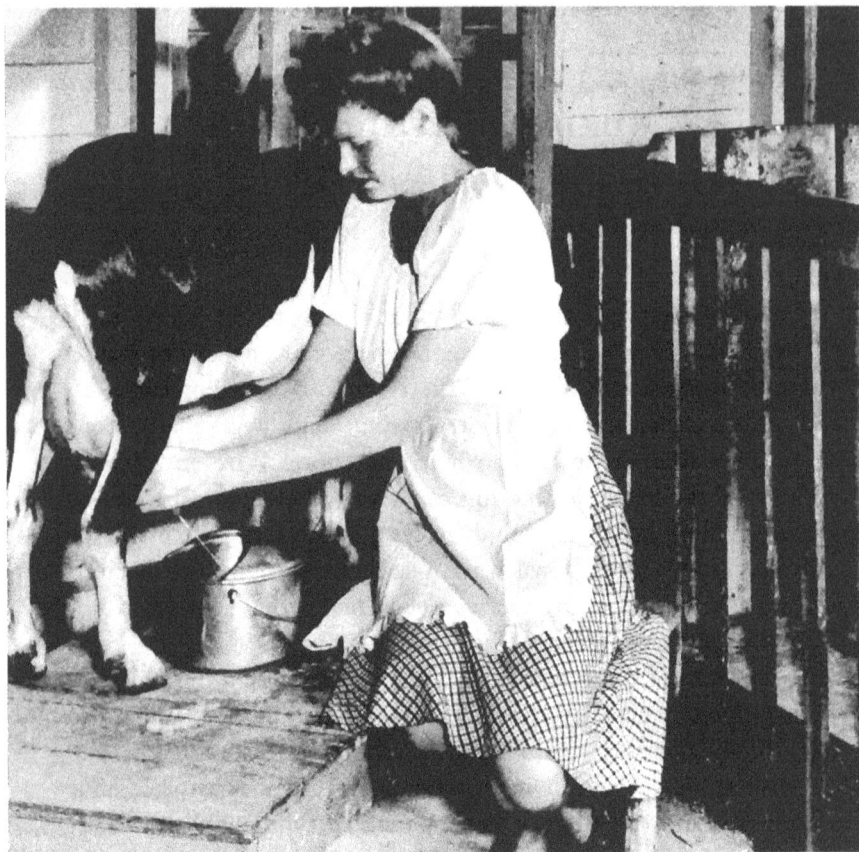

Milking a goat is not difficult. It is best done in a separate room. Note milk stand and covered pail.

the udder and such milk should not be used until the condition has been remedied.

As a check on yourself it is a good idea to have the milk tested from time to time by your local board of health. Bring in a pint and they will send you a report of the bacteria count, the amount of butter fat, and the sediment, if any. The cost will probably be nothing, or too little to worry about. If the milk is below standard, the remedy is in your hands.

For the sake of cleanliness and comfort your goats should be milked on a bench, and does coming from a dairy herd have

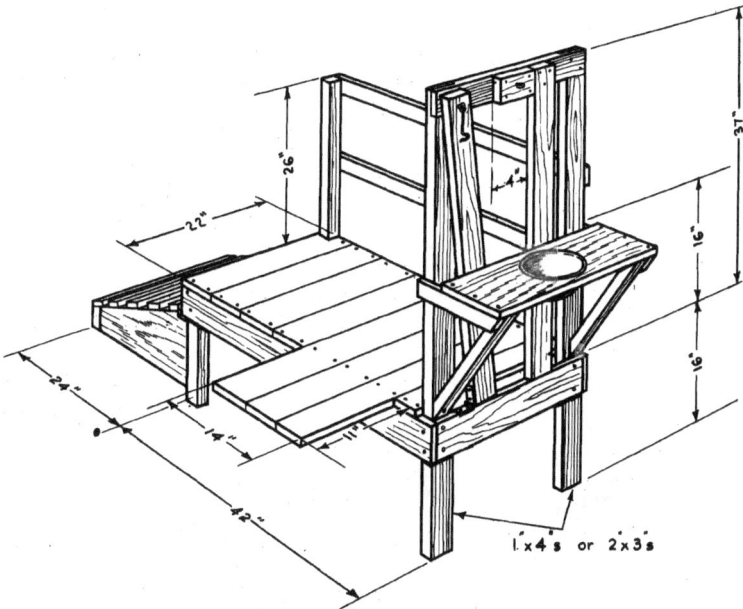

Details of milking stand for goat.

been trained to this routine. If your doe is a first milker or has not been accustomed to the bench you may need help to boost her onto the stand for the first milking or two, but she will learn readily. You may find it advisable to have a riser with a cleated surface so that the goat can walk up onto the stand, although most goats seem to like to jump up. It is all a matter of early training and the best plan is to continue in the way to which she has been accustomed.

Have a shelf beside the bench and your equipment right at your elbow so that once you have started to milk you don't have to hop up for one thing or another. Brush the goat first, and don't neglect the under part of her body or the inner part of her hind legs. Wipe off her udder with a cloth wrung out of warm water to which you have added a few drops of chlorine solution, and dry her udder and your hands before proceeding

Folding milking stand—open and closed.

to milk. A handy roll of paper towels is best for this—a clean, fresh towel for each goat. Be sure that your hands are clean, and be careful not to touch the inside of the containers, otherwise your careful sterilizing will serve no purpose.

It is also well to keep in the barn a cotton frock with long sleeves such as grocers use which can be slipped on before the milking. Some health boards require that a band, or apron, be fixed about the goat's body to prevent hair from falling into the milk. You can make a simple apron from half a cotton grain bag without sewing a stitch. Just make a cut half way down each long side, about an inch or two in from the edge, tear a strip along the edge to about an inch from the end and tie a knot to prevent further tearing. This gives four tie strings. Pass two under the front legs and tie them across the goat's back; pass the other two under the hind legs and tie them across her back, over the hips.

If you can milk a cow, milking a goat will be easy for you, but if you have never milked before you will wonder how in the world with such a large container under the goat you can squirt the milk in so many different directions. The man from whom I purchased my goats supervised the making of the milking stand which had a rectangular platform on which the goat stood with head locked in place. I sat on the side and by the time I had finished milking there was milk up my sleeves, on my face and on the walls, and my legs ached from the strained position so that I could hardly keep from tears. The next time I tried sitting on a stool beside the bench. This was better, but too far away from the goat. Then came the happy thought of incorporating the seat with the platform. This was as it should have been from the start, but mine was the trial and error method. From then on the milking was no longer a torturing process and my aim was much surer. Once I had seen an Italian out in the pasture milk directly into a bottle, which seemed almost like magic, but soon I too could hit the bull's eye.

Here's how to milk: Sit on the milking stand facing the goat's udder, your shoulder close to her shoulder.

Encircle the teats, your thumbs outside your fingers. Close your grasp, beginning at the top (thumb and index finger) and successively close the other fingers, thus forcing the milk down the teat and out in a steady stream. You will note that if you don't shut off the milk with your thumb and index finger, the milk will flow back into the udder when you squeeze with your second, third and little fingers. The motion of the fingers in milking is somewhat similar to playing the scale on the piano, except that the fingers are kept close together. That is the proper way to do it, but if you find it too difficult to operate each finger separately, pressing the teat with all the fingers at once will do.

The teats should be grasped and pressed alternately—not the udder—with as little motion as possible in the upper arm. The pressure should be firm but gentle. Pulling and stretching the udder may cause injury to delicate blood vessels. If the doe gets restless and moves her hind leg, protect the pail by pressure of your arm against her leg, and hold to the teat. If she moves her foreleg, pressure against it and against the under part of her body will control this. Sometimes a first freshener who has been nursing her kids will lie down on the bench to prevent being milked. A band passed under her forelegs and fastened over the front of the stanchion will keep her on her feet.

If your goat has been accustomed to being milked on the side opposite the side at which you milk, be patient with her until she becomes adjusted to the new position. At one time a doe was left with me for a week. Each time she was milked she tried to lie down. When her owner came I asked about this. He was puzzled. We put her on the bench to be milked and sure enough, down she went, even with her owner milking her. Then the light dawned and I asked, "On what side do you milk?" "The right," he said, and my bench was for left side milking.

Occasionally, particularly among Saanens, a young doe even before breeding, will develop milk in her udder—a virgin milker. If it seems necessary to relieve the pressure, she should be milked regularly.

When you think you have milked the goat out thoroughly, nudge her udder a few times and more milk will come, and lastly run your forefinger and thumb down the teat (this is called stripping) until you have gotten out the last drop. This isn't just Scotch thrift, but is necessary for thorough milking and keeps up the doe's production. Then, too, the milk that comes with stripping is the richest milk in her udder.

As soon as you have milked out the doe, pour the milk through the strainer, even before you return the milker to her stall, and when the milking is completed set the container in cold water—ice water if possible—as much water as you can use without tipping over the container. To ensure good flavor in milk and prevent the development of bacteria, rapid cooling is important, and cooling by immersion in cold water is a more rapid process than dry refrigeration. The milk should be

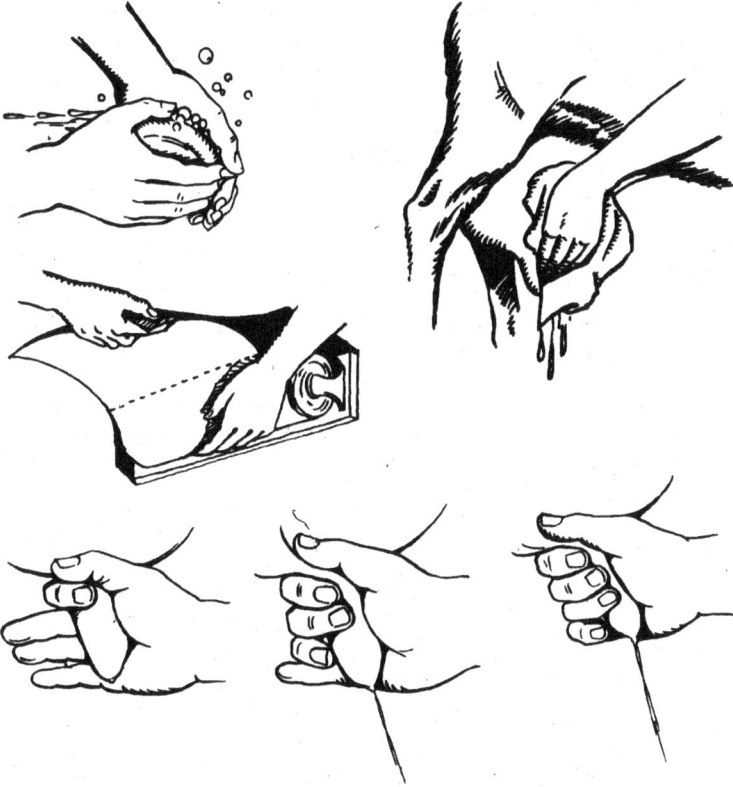

*Upper left.* In milking follow the same routine at each milking. Be gentle. Strangers—and the family dog tend to make a doe "hold up her milk." Keep them out of the barn or milk room. After milk pail and wash pail are ready, be sure your hands are clean. *Upper right.* Wash the goat's udder in chlorinated warm water—from 120°– 130° F. The udder should be washed *just before milking.* Use a separate wash rag for each goat. *Center left.* Have a roll of paper towels handy to milking stand. Dry the goat's udder and your hands. Wet hands can cause a chapped udder—and worse. *Lower left.* Milk can run out of the teat into the pail or *back into the udder.* So first close your thumb and first finger so the milk can not run back into the udder. *Lower center.* Next close your second finger— and the milk should squirt out. Discard the first stream—it will be high in bacteria. *Lower right.* Close the third finger. Use a steady pressure. Don't jerk down.

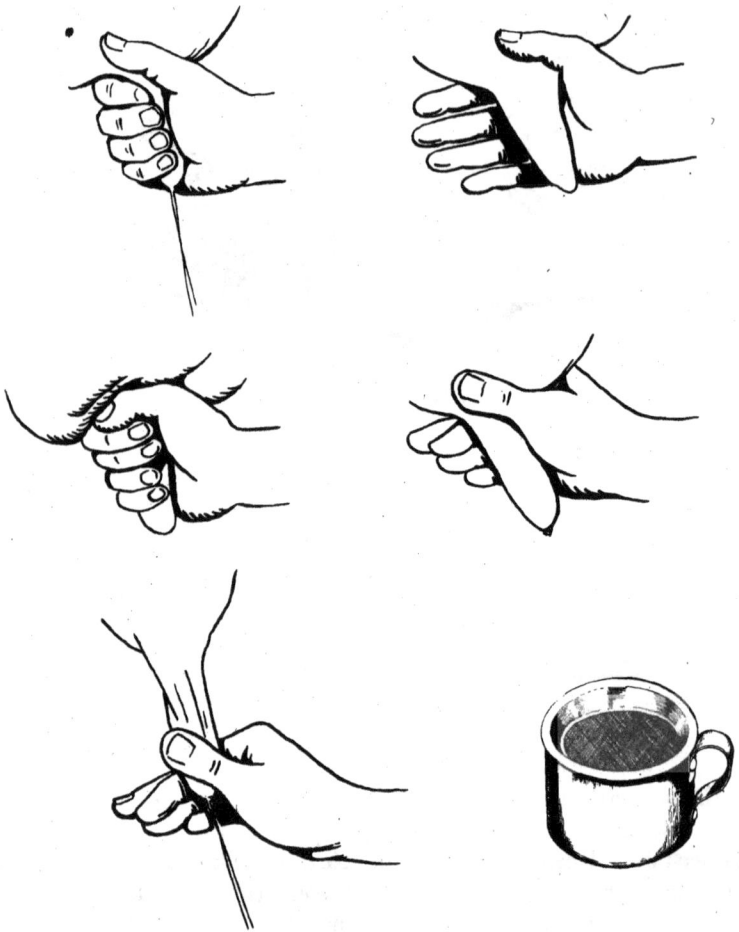

*Upper left.* Next close your little finger . . . squeeze with whole hand. *Upper right.* Now release the teat and let it fill up with milk. Repeat the process with the other hand . . . *Center left.* When the milk flow is near to stopping, nudge the bag to see if the doe has let down all her milk. *Center right.* The final bit of milk may be stripped out. Take teat between thumb and first finger. *Lower left.* Now run down length of teat. Milk high in butter fat usually comes at end of milking. But prolonged stripping is bad for the teats and udder. *Lower right.* Strip cup: the first milk is milked into the strip cup. If the milk is "lumpy" it will not pass through the strainer.

brought down to a temperature of 50° F. within one hour after milking. Cubes from the refrigerator ice tray are fine to use, or if no ice is available change the water frequently as the warm milk will quickly raise its temperature.

When the milk is cooled it may be transferred to bottles, capped and placed in the refrigerator. Gallon glass jars (such as are used for mayonnaise), or quart canning jars make excellent milk containers and are much easier to clean than the customary milk bottle. As the milk is used the metal cap which screws on the outside of the jar can be removed and replaced with less danger of contaminating the milk than with the reused bottle cap made of cardboard. The lining, of course, should be removed from the cap, and a piece of waxed paper may be used in its place to protect the milk from splashing against the metal cover.

All milking equipment should be rinsed in cold water immediately after use then washed in warm, soapy water to which chlorine solution is added. This solution is available at your grain dealer's, or you can make it yourself by dissolving one 12-ounce can of chloride of lime in a gallon of cold water. After the lime has settled pour the water into bottles. Add this solution to your wash water in the proportion of 4 tablespoons of the chlorine to 3 gallons of water. Finally the utensils should be rinsed in boiling water and set in a dust free place to dry by evaporation. As a further precaution this last step may be repeated just prior to milking for the extremely hot water will dry the utensils very quickly.

Don't leave milk to sun itself on the kitchen table, but be sure that it is kept always in the refrigerator, properly capped. Don't set a pitcher of milk on the shelf uncovered and then blame your goat for off-flavored milk if it doesn't taste good to you. Raw milk absorbs flavors very quickly, and must at all times be kept covered.

# Goat Milk and Cream

MILK as it comes from the animal, goat or cow, is referred to as raw milk. Very many people whose animals are tested every six months for TB and Bang's disease have no hesitation about using the raw milk if the animals show a clean bill of health, which is usually the case with goats. These people maintain that pasteurization affects the flavor and the calcium and vitamin content of the milk and they prefer it raw. Others, however, feel safer if the milk has been boiled or pasteurized. In some communities the health regulations require that dairies selling milk commercially must have the milk pasteurized, for pasteurization kills the bacteria that cause TB and undulant fever in humans.

Pasteurization* may be done at home by heating the milk

* "The Effect of Pasteurization on some Constituents and Properties of Goat Milk" Haller, Babcock and Ellis, *U. S. Department of Agriculture Technical Bulletin 800*, says:
"The milk was pasteurized by holding it for 30 minutes at a temperature of from 142-147°F., or by bringing it to 160°F. for 15 seconds. The solubility of calcium (lime) and phosphorus was only slightly decreased, to an extent less than the normal variation in composition. The protein was not appreciably affected, but the curd tension was considerably reduced by the 30-minute method and only slightly reduced by the 15-second method. The flavor is said to have been slightly improved and the keeping qualities considerably improved. The phosphates test which is usually applied to cow milk to determine the efficiency of pasteurization could not be applied to goat milk as the phosphate enzyme is destroyed very rapidly. The vitamin C content was reduced about 40 per cent by the 30-minute method and not at all by the 15-second method."

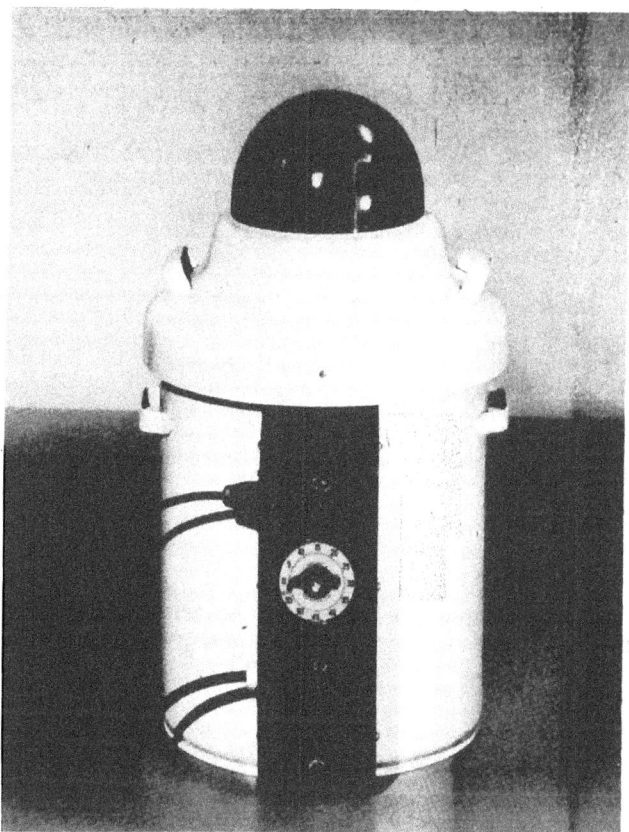

This is an electric, home-size pasteurizer ideal for small dairies; it sells for around $40. It makes pasteurizing easy and eliminates the chance of getting TB or Undulant Fever.

to at least 143°F. and holding it at this temperature for at least 30 minutes, then cooling it rapidly; or by heating it to at least 160°F. and holding it at this temperature for at least 15 seconds, and cooling rapidly. Use a dairy thermometer for this—it must not be done by guesswork and be sure that every drop of the milk receives the heating. To keep the milk pure, even after pasteurization, it must have the same careful handling and refrigeration as in its raw state.

In *Technical Bulletin #46* of the New York Agricultural Experiment Station the following figures are given on the comparative food value of Goat Milk, Cow Milk and Human Milk.

|  | Fat | Milk Sugar | Protein combined with Calcium | Salts |
|---|---|---|---|---|
| Goat Milk | 3.80 | 4.50 | 3.10 | .939 |
| Cow Milk | 3.90 | 4.90 | 3.20 | .901 |
| Human Milk | 3.30 | 6.50 | 1.50 | .313 |

*Per cent*

A similar analysis in the book, "Milk and Milk Products," by Eckles, Combs and Macy gives the following:

|  | Fat | Lactose | Protein | Ash | Water |
|---|---|---|---|---|---|
| Goat Milk | 3.82 | 4.54 | 3.21 | .55 | 87.88 |
| Cow Milk | 3.80 | 4.80 | 3.50 | .65 | 87.25 |
| Human Milk | 3.11 | 7.18 | 1.19 | .21 | 88.30 |

*Per cent*

From these figures it is notable that while the constituents of goat milk and cow milk do not differ greatly in percentage, human milk shows a lower percentage of protein and a higher percentage of lactose, or milk sugar, than either goat or cow milk. In fact in the Milk and Milk Products table which lists also the analysis of the milk of the sheep, mare, water buffalo, reindeer, camel, sow, bitch, and cat, human milk is lowest in protein and highest in lactose of all the milk analyzed.

The great value which goat milk possesses lies principally in the fact that it forms small, soft curds in the stomach and that the fat globules are small and well emulsified, which makes the milk easily and quickly digested. For this reason it is often advised by physicians for infant feeding, although usually in modified form for the infants, with the addition of malt sugar, or honey, and boiled one minute or pasteurized. Because of its delicacy and easy digestibility it is especially beneficial for invalids or people with low vitality or sensitive digestive systems. Also, in many instances it can be tolerated by people allergic to the protein of cow milk, especially when the allergy results in eczema, hay fever or asthma.

A neighbor of mine occasionally stopped in for a small bottle of goat milk, much as one would stop at a soda fountain. Then one day she explained that she had been using the milk in her nostrils for an irritation of the mucous membrane. She insisted that the milk had cured the irritation. This was too far-fetched for my credulity, and I thought of the scripture quotation, "Thy faith hath made thee whole." Who knows? Cleopatra took her milk baths—probably of goat milk—why? If it was just to be clean she could have found plenty of water in the Nile. There are countless cases of people with one ailment or another who give to goat milk credit for their recovery. One trade publication carried an article by a man who put himself on a goat milk diet to cure pyorrhea and claimed that it was successful. Some of these "cures" even though unscientific are not unworthy of consideration when they come from reliable sources. However, goat milk need not be regarded as a nostrum, or a cure-all intended only for the sick or undernourished, and it isn't wisdom to put oneself on a diet of any milk, even goat milk, exclusively. According to scientific authorities, while milk is an excellent food for adults as well as for children, the amount of milk needed for an adult limited to milk alone would be eight to eleven pounds (about four to five and a half quarts) a day, and the protein content is too high and the solid material too concentrated for the average adult (Eckles, Combs and Macy).

Goat milk has its place as a healthful, energy producing food and a delicious beverage for the sick and the well. I know of no milk so refreshing or equal in flavor to a glass of cold, clean, raw goat milk. If you tire of it occasionally as is, try ginger ale with it, fifty-fifty. It is a real thirst quencher on a hot day. Or make iced coffee and use the whole milk in it, not just the cream. Even a gourmet would enjoy it.

The Agricultural College Experiment Station of New Mexico in *Bulletin #154* reports an interesting test regarding the flavor of goat milk as compared with cow milk. A sample of each was given to twelve persons, men and women. No explanation was made and no indication given them that the milk was from animals of different species. It was just milk. They were asked which they preferred. Seven of the twelve chose the goat milk. A similar test was made with fourteen students at State College, New Mexico. Eight preferred the goat milk, four cow milk and two had no preference.

One sometimes meets the comment "but you can't get cream from goat milk!" You can, indeed, although because of the small fat globules and the more complete emulsification it takes longer for cream to rise on goat milk than on cow milk, unless a cream separator is used. With a small separator, however, the goat milk separates just as quickly as the cow milk.

You can make butter from goat milk; also cheese; ice cream from the cream and sherbet from the whole milk; and in the preparation or cooking of any food requiring milk, goat milk functions just as adequately as cow milk. Try cream soups or clam chowder or oyster stew made with goat milk and prove it yourself.

# Making Butter at Home

AT ONE TIME I visited the owner of a large herd of goats, and as we talked the man who did the milking appeared with a pail of milk—gallons of it—that he poured into a galvanized tub for three or four large dogs to drink. I shrank a little mentally, almost as though the milk had been tossed into my face, for I have a high regard for goat milk as human food and could think of other dishes more suitable for feeding grown dogs. The owner remarked that they had more milk than they could use, and I wondered why they didn't make butter. To be sure it is some trouble, but no more than the making of butter from cow milk.

There are several methods for extracting the cream. Many home buttermakers still cling fondly to the old-fashioned "shallow pan" method. The milk is placed in flat pans and set in a cool place until the cream rises. The cream is then taken off with a "skimmer"—a slightly concave disc with perforations through which the skim milk drains back as the cream is lifted out. The trouble with this is that despite your best efforts there is always danger of dust and bacteria getting into the milk.

There is another method, the "deep setting" method in which the milk while still warm is poured into a "shot-gun" can immersed in cold water (preferably ice water). The cream is skimmed off, or poured off unless the can is equipped with a spigot for drawing off the cream and skim milk separately. This method takes about half as long as the shallow pan method.

The most modern method, however, and by far the best, is the centrifugal separator. With this method the warm milk (which should be 80 to 90° F.) is poured into the separator. This whirls it around and in a jiffy out comes the cream through one spout, the skim milk through another. This is especially desirable for separating goat milk, as goat cream, because of its greater emulsification, rises more slowly than the cream of cow milk, and some goat milk as it ages develops a cheesy or "goaty" flavor which makes the butter unpalatable.

If buttermaking is to be included in your accomplishments, a separator is a good investment. The cost is between $20 and $30 for a small, table model, and it is one of those things that proves its value in the greater efficiency with which it accomplishes your purpose, not from the viewpoint of "Do I get my money back through its use?"

*Cornell Extension Bulletin #269*, quoting from *Bulletin #116* of the Indiana Agricultural Experiment Station, has a table that gives the figures showing the amount of butterfat retained in skim milk in various methods of separation:

| Method | Percentage of Milk Fat |
|---|---|
| Modern | 0.02 |
| Deep-setting | 0.17 |
| Shallow pan | 0.44 |
| Water dilution | 0.68 |

The "water dilution" method, in which water is added to the milk to aid in creaming, is considered least satisfactory. It means loss of butterfat, water-flavored butter, and diluted, watery skim milk, unsuitable for cheese making or other household use and even for animal feeding.

Installing a separator, however, is no easy job for a woman unless she has a flair for the mechanical. It has many tricky little pieces—set screws, discs, etc., and must be set in place with the guidance of a spirit level, but the instructions for putting it together are simple in man's language, and almost any man can cope with them successfully.

If you purchase a separator have it installed in a cool, clean place and be sure that it is properly balanced and firmly set. Otherwise you will run into difficulty, for the machine is delicately constructed and must be protected against uneven wear on its carefully adjusted parts. Sterilize it each time after use, just as you would any other utensil used in buttermaking. This is important, as stale milk in the separator means fertile soil for bacteria.

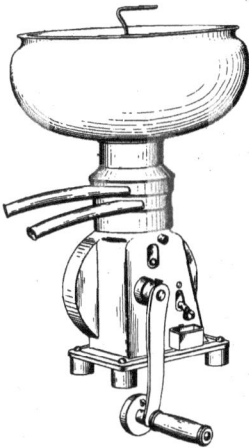

This small separator sells for around $30.

### COOLING THE CREAM

After the milk is separated, place the cream immediately in cold water (not in the refrigerator). Water will cool it more rapidly and uniformly than cold air. Mix it occasionally with an up-and-down motion. There is a special "stirring rod" made for this purpose, or you can use an ordinary perforated vegetable ladle with a long handle. When cooled, place the cream in the refrigerator until you are ready to use it.

### PREPARING THE CHURN

Small churns for home buttermaking may be secured from dairy supply houses or mail order houses at prices ranging from $2 to $5. If your churn is of glass it needs the same care as your milking equipment—immediate washing after use and sterilizing before the butter is made, followed by rinsing in cold water. A wooden churn should be scalded with boiling water, then chilled with cold water before use. If you make butter only occasionally, fill the churn with hot water at least twenty-four hours before you churn so that the wood will expand (otherwise it may leak) and then be sure that you rinse it with cold water.

For small amounts of butter many people find the electric cake mixer a very satisfactory churn.

A gallon of cream will make two or three pounds of butter and in using goat milk it may be necessary to hold the cream from several milkings in the refrigerator before there is enough to work with. Each lot must be kept separate and none of it should be more than three or four days old. Old cream makes inferior butter that deteriorates rapidly unless pasteurized.

About twelve hours before you plan to churn put all the cream together and mix it thoroughly with your stirring rod to give it uniform thickness.

When you are ready to make the butter warm the cream slowly to 65–75°F. Use a dairy thermometer for this (you can purchase one for about a dollar). Mix the cream frequently with the stirring rod or ladle, and keep it at this temperature until it becomes thick and glassy and tastes a little sour.

Reduce the temperature of the thickened cream rapidly to 52–60°F. in the summer, 58–68°F. for butter made in the cold weather, and hold it at this point for at least two hours.

In pouring the cream into the churn pass it through a strainer so that any lumps will be broken up, and have the churn ⅓ full of cream to prevent overcrowding as the butter expands. Goat milk makes white butter and always needs a few drops of vegetable coloring to give it the shade you like (10 to 20 drops per pound of finished butter).

After the cream and coloring have been placed in the churn turn the handle about ten times, stop and remove the plug or lift the cover of the churn to let the gas escape. Replace the plug, turn the handle about twenty times and again let out the gas. Continue churning until the butter granules are about the size of a pea. Start churning at a speed which will produce the great-

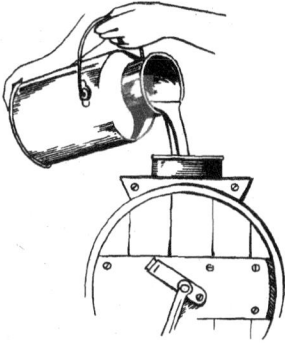

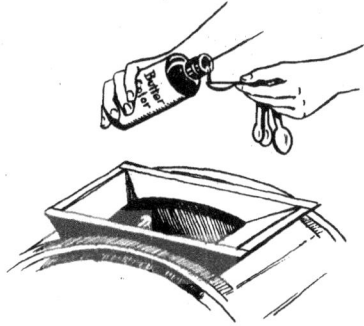

*Left.* In churning butter cream is heated to 52–60° F. in summer; 58–66° F. in winter, then strained into clean churn. *Right.* Churn should be only ⅓ to ½ full. Next butter color is added.

est concussion which can be determined by the sound—about 60 revolutions per minute for the common barrel type of churn. Too fast churning makes cream cling to the one end rather than fall from one end of churn to the other as it should. As you become adept at buttermaking the change in the churning sound from a swish to a heavier sound will tell you how you are progressing. In thirty or forty minutes the butter should be formed.

### WASHING THE BUTTER

Drain off the buttermilk at this point through the strainer to rescue any butter particles in the milk, and pour water of the same temperature as the buttermilk over the granules. Have about as much water as there is buttermilk. Close the churn and turn it a few times rapidly to wash the butter. Draw this water off, and pour about the same amount of fresh water, at the same temperature, into the churn. If the butter is too soft use colder water for this rinsing, if too hard have the water a little warmer. This water when drained off should come pretty clear. It is important that all the milk be washed out otherwise the butter will develop an unpleasant flavor as the milk proteins deteriorate.

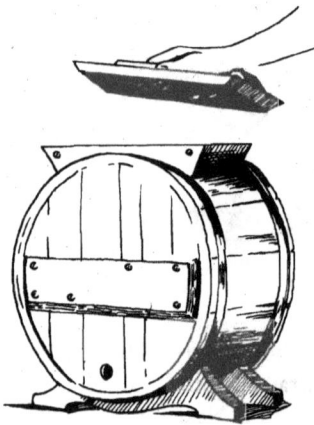

*Left.* Churn is turned at about 60 revolutions a minute. After a few seconds, remove cover and let gas escape. Repeat 2 or 3 times during early stages. *Right.* Churn until granules are size of wheat grains. Let out buttermilk through strainer.

*Left.* Wash butter in twice the amount of clean water as there was buttermilk; water should be same temperature as buttermilk. Add ½ water to butter in churn, turn four times, drain—then repeat. *Right.* Salt (¾ ounce per pound of butter.) Press butter into a thin layer, then fold into pile and press again. Continue until salt has even distribution and butter has good body.

## WORKING THE BUTTER

Remove the butter to your "worker" if you have one, or to a wooden bowl or wooden tray which has been rinsed in cold water. Sift good quality table salt, ¾ ounce for each pound of butter, over the surface and press the butter into a thin layer with a wooden paddle. Fold it over and press it again. Be careful to press, not smear the butter. Smearing makes it like a salve. When it is firm, close-grained and waxy you have finished and can enjoy your toast.

## CARE OF EQUIPMENT

Wash all buttermaking equipment thoroughly before putting it away, with hot, soapy water, rinse it thoroughly and store it where the wooden parts won't become bone dry.

# Making Cheese From Goat Milk

EXCEPT for some methods of making soft or cottage cheese, all cheese is made by coagulating milk with rennet. It is desirable, however, to hasten the coagulation by adding a starter—which is milk previously coagulated or clabbered by souring. Even in making cottage cheese, the best is made by the additional use of "Junket" rennet tablets, which not only hasten the coagulation, but make a more desirable "sweet curd" cottage cheese. For hard cheese—American cheddar, etc.—only a small amount of starter and more rennet are used; this is necessary in order to produce cheese of the desired texture, the kind that develops in flavor as it ages and should be weeks or months old before being consumed.

Cheese may be made from skim milk, whole milk, or from cream, from whey and from buttermilk.

Cheesemaking is a subtle process, for in addition to the lactic acid bacteria which are propagated through the heat and the addition of rennet or a starter, there are other undesirable bacteria in the milk which become active under the heat. These must be kept at the lowest possible minimum. Your strongest weapon against them is cleanliness—the first and most vitally important factor involved in the making of cheese. Animals must be clean and healthy, kept in clean quarters and the milking done in a clean, dust free place. All milking utensils and all equipment used must be sterilized. This includes knives, spoons, the kettle into which the milk is poured, cheesecloth and muslin drain cloth, the thermometer—every item used in the cheese-making.

The second factor is careful observance of directions, especially with respect to temperature control. Hit or miss guess work will not bring good results.

The third and most difficult factor, involved in making cured or ripened cheese, is the conditions under which the cheese is ripened. Different varieties of cheese, of which there are many, require different conditions of temperature and humidity during the curing period. Before you make a specific kind, be sure that you can meet these conditions. A cheese that starts out with every promise of success, carefully and cleanly made, may through variations in temperature and humidity while curing become greasy and rancid and unfit for use. And remember, too, that cured cheese takes time and patience to make. If you're a person who wants results in a jiffy, don't undertake it.

The best place for curing cheese of the hard type is a root cellar, a spring house, a cool, old-fashioned pantry, or the house cellar if it isn't damp. The temperature should be maintained at 50 to 55°F. and the humidity not more than 85 per cent.

In cheesemaking perhaps more than in most fields, experience is truly the best teacher, and with repeated efforts you learn to recognize the appearance and the feel of the batch that holds promise. You will come to recognize and prevent some of the commoner defects such as:

Sour or acid cheese which results from the use of old milk or from too much whey left in the curd before draining. This, when due to the latter condition, can be avoided by raising the cooking temperature, or by cutting the curd into smaller pieces so that the whey drains off more freely.

Bitter flavor, due to bacteria in the milk. This will not occur if milking conditions and handling are sanitary.

Sweet or fruit flavor also may be due to uncleanliness or to improper development of acidity during making.

279

Coarse texture, caused by insufficient pressure, or pressure at too low temperature.

Cracks caused by improper bandaging or insufficient pressure, can be corrected by dipping cheese into warm water to soften rind and pressing again.

During the summer months when milking goats are on pasture they are at their best production and you may have surplus milk for making cheese. Most housewives with an extra quart of milk have made soft cheese now and then with good results either from the whole milk or from skim milk after the cream has been separated for coffee or cereal. It is very easy to make and takes only a little time, but soft cheese has keeping quality of only two or three days and too much should not be made at one time. A gallon of milk will make approximately 1½ pounds of cheese.

## Cottage Cheese

The skim milk is allowed to sour naturally at a temperature of 75°F. It can be set on the radiator or the back of the kitchen stove and takes about 24 hours to thicken or clabber.

When it is firm and smooth with a little whey on top it is cut into ¾" cubes and the container placed in another larger container, holding water of the same temperature—75°F. Very gradually the temperature is raised to 100–110°F., not more than 1 or 2 degrees in five minutes, and the contents stirred gently as it heats, to prevent the curd from becoming lumpy. It is held at this temperature for fifteen to thirty minutes. When a small piece pressed between the fingers holds together and shows no milky leakage the curd is transferred into a sack or drain cloth and hung in a cool place to drain. When cool, salt is mixed gently into the curd, about one teaspoonful to one pound of cheese.

## Making Starter

You can make starter from fresh whole milk by placing a small amount—about a pint—in a sterilized jar and setting it aside to sour at a temperature of 70–75°F. Put waxed paper over the top of the jar to keep out dust. When soured the milk should be smooth and free from gas holes, with a sour odor and flavor. Remove the top with a sterilized spoon, cover the jar and set it in the refrigerator until needed.

Before using the starter, mix it evenly by transferring it to another sterilized jar and back to the first one.

The following recipe is for cottage cheese made with starter and ⅛ "Junket" rennet tablet combined:

> 1 gal. skim milk warmed to 75°F.
> ¾ cup starter
> ⅛ "Junket" rennet tablet dissolved in 4 tbsp. cold water

Add starter to milk, then add ⅛ dissolved "Junket" rennet tablet, stirring thoroughly.

Hold milk at 75°F. until smooth, firm curd has formed.

Ladle, without cutting, into sack and drain.

When practically free of whey add salt to taste.

A little cream added to cottage cheese just before serving gives it richness and flavor. It is a very versatile dish and during the warm weather particularly can be made the basis of many delicious salads, with the addition of pimento, chives, chopped radish, sweet peppers or other vegetables. Fruits such as pineapple, orange, grapes or cherries can be worked into delicious, healthful combinations with cottage cheese, or for the children the sweeter fruits like raisins, cooked prunes, peaches or pears have a strong appeal. Experiment with it and find your own favorite combinations.

Best of all, cottage cheese is easily digested and highly nutritious. The Vermont Agricultural Extension Service gives figures showing the relative protein content of cottage cheese as compared with various meats and fish that may surprise you.

One pound of cottage cheese equals in protein:

| | | |
|---|---|---|
| Beef Rump | 1.31 | lbs. |
| Leg of Lamb | 1.27 | " |
| Beef Rib Roast | 1.55 | " |
| Fowl | 1.52 | " |
| Halibut | 1.36 | " |
| Salmon | 1.51 | " |
| Average | 1.36 | " |

From this average it is evident that the cottage cheese contains one-third more protein than the meat and fish, and when this cheese is made in the home from surplus milk the cost is next to nothing.

### Cottage Cheese Omelet

Here are a few simple uses for cottage cheese with fruit:

Cut raisins into pieces and mix with cheese; place on lettuce leaves; top with cherry. Serve with mayonnaise flavored with pineapple juice.

Mold cottage cheese into ball; surround with unstrained cooked cranberries; or mix cranberries into cheese if desired. Serve on lettuce.

The U.S. Department of Agriculture suggests these cottage cheese dishes:

### Cottage Cheese Omelet

| | |
|---|---|
| 3 rounded tbsp. cottage cheese | 1 tbsp. chopped pimentos |
| 2 eggs | 2 tbsp. milk |
| ¼ tsp. salt | ⅛ tsp. soda |

Beat yolks and whites of eggs separately. Add to yolks the salt, milk and cheese with which pimentos have been blended. Fold in stiffly beaten whites. Pour into hot frying pan in which ½ tbsp. fat has been melted. Cook slowly until egg has set; place in oven to complete cooking; fold in center. Garnish with parsley.

### Cottage Cheese and Pimento Roast

2 cupfuls cooked lima beans
3 canned pimentos, chopped
¼ lb. cottage cheese,
bread crumbs, salt

Put beans, pimentos and cheese through meat chopper. Mix thoroughly. Add salt. Add bread crumbs until stiff enough to roll. Brown in oven, basting with fat or butter, and water.

### Cottage Cheese and Nut Roast

1 cup cottage cheese
1 cup bread crumbs
1 tbsp. butter, (salt, pepper)
1 tbsp. chopped walnuts
2 tbsp. chopped onion
juice of ½ lemon

Cook onion in butter and small amount of water until tender. Mix remaining ingredients, moistening with water in which onion has been cooked. Place in shallow baking pan and brown in oven.

### Another Soft Cheese, Neufchatel

The French make a soft cheese, Neufchatel, from whole milk, similar in appearance to cottage cheese but, of course, much richer. A delicious cheese of this type requires:

1 gal. sweet, whole milk, tempered to 70°F.
1 "Junket" rennet tablet dissolved in ¼ cup cold water

Place milk in kettle set in larger kettle containing water of same temperature as the milk, 72°F. The milk is ready for one "Junket" rennet tablet dissolved in ¼ cup cold water and stirred in thoroughly.

If this is done in the evening, leave undisturbed until morning when the curd should be smooth and firm with very little whey on top. This takes about twelve to fifteen hours.

Ladle the curd into a drain cloth spread over the colander and when fairly dry tie the ends of the cloth and press the curd under a board with a weight on top.

Stir it occasionally to help this whey to drain off.

When dry, add salt to taste.

To ensure a good flavor drain and press the curd in a cool place.

This is delicious. A native Frenchwoman who tried mine immediately wanted to buy some.

### Buttermilk Cheese

At the dairy where I went for buttermilk the man said he had never heard of making cheese from it. A neighbor who made butter also had never heard of the cheese. I tried several recipes and got best results from the following:

Heat sour buttermilk slowly to 95°F. Cover and leave undisturbed for about one and a half hours. Then raise temperature very slowly to 140°F. and hold at this temperature until the curd sinks. Drain and salt to taste.

This makes a delicate, soft cheese, fine textured and creamy.

### Processed Cheese

This cheese is a little more complicated, but not difficult to make and well worth the extra work.

10 qts. thickly clabbered milk
4 tbsp. butter
¾ tsp. soda
1½ tsp. salt
⅔ cup sour cream
⅛ cheese color tablet if desired.

Heat milk gradually to 125°F., stirring occasionally.

Ladle into drain sack and hang until dry, mixing curd occasionally to hasten draining.

When dry add 4 tbsp. butter, ¾ tsp. soda, 1½ tsp. salt, mixing it in gently.

Press down in bowl and set in warm place for two and a half hours.

Place curd in double boiler, add ⅔ cup sour cream, ⅛ tsp. cheese coloring dissolved in ¼ cup water and heat until liquid.

Pour into jars and seal.

### Mysost

Every time I made cheese and saw the clear, pale greenish whey disappearing down the drain I felt guilty. I knew that it had food value and it seemed unpardonable to waste any. I had no pig and no chickens, but put out a pan for Chester and Mabel, the ducks. They washed their bills in it, but as food it had no interest for them. I set out a sample for the cat. She took a few sips and walked away. Some of the older goats drank it but next day had a tendency toward scours. As the young kids were on a strictly whole milk diet it wasn't safe to give them any.

I read about whey and found that it contains most of the lactose or sugar of the milk, minerals and albumin. And I learned that among Scandinavian people this whey is boiled down and made into primost or mysost. So I tried it.

Strain the whey and boil it down rapidly, stirring almost constantly. Skim off the albumin that rises to the top, and when the whey is reduced to about one-quarter the original amount put this albumin back and stir it in thoroughly. Continue boiling until it becomes thick. Pour quickly into a wooden bowl and stir it until cool to prevent sugar crystals from forming.

The boiling down process is a lengthy one and the almost constant stirring necessary to prevent scorching becomes a bit tedious, but the fruit of your effort is an attractive spread about the color of maple sugar, sweet and bland that some people find delicious.

#### HARD CHEESES

From time to time I looked over recipes for cured or ripened cheese. The array of figures and the hoops and molds mentioned frightened me off. I got the impression that hard cheese-making was involved and difficult. Then a friend told me that he had a simple recipe given him by a Frenchman and he would show me how the cheese was made.

The recipe *was* simple and easy to follow, but like the old lady who marked her pies "TM"—" 'tis Mince" and " 't'aint Mince"—he couldn't tell with any sureness which of his cheeses were good and which were not good, and they all had quite a coating of colorful mold, which meant waste. The sample I had was a " 'tis good" and I decided to try the recipe.

Coffee cans such as he used for molds were hard to find as coffee was being packed in jars and paper bags, but finally I located two bright, new ones and punctured the bottoms with holes to allow the whey to drain off. Then I made little wooden discs, a shade smaller than the circumference of the cans, called "followers" in the cheese business.

The cheese looked right and the curd was dipped into the cans which were lined with cheesecloth, the little wooden followers placed on top, and for weight a quart bottle of water set on top of each cheese. During the night I heard a crash in the kitchen and rushed out to stalk the burglar. I found that the pressure of the curd had pushed the bottles off the cheese, spilling the water. The resilience of that curd astonished me. Flashlight in hand, I hurried out to the stone wall and groped about for two smooth, round stones which I scrubbed and placed on the cheese for weights. Next day when the cheese was removed from the molds the edges had to be trimmed and the top surface smoothed off where the folds of the cheesecloth had left ridges—more waste.

My friend cured his cheese in a root cellar. I had none and decided that the stone garage would do, but in a few days mold developed. This I wiped off with warm salt water and transferred the cheeses to the milk room. A nice yellow rind began to form. Then there came a week of stifling hot weather. The cheeses sweated, grew hard and rancid. Much discouraged I tossed one out to my neighbor's cat. She struggled with it, holding it down with her paw, but even her sharp teeth could make no impression on the tough rind and she gave it up. I gathered up all the cheeses, about a half dozen, concluded that with cured cheese I was a failure, and put them in the garbage can. Later

I learned that the well is a very good place for curing cheese made during hot weather.

Then I came upon the Hansen chart for making hard cheese; it looked so simple that I tried again. Over a period of several weeks I made two cheeses a week. When they were cured a neighbor bought one at $1.25 a pound, and within a half-hour brought in a friend who wanted one to take back to the city. Another cheese I kept in an electric refrigerator for a year. No mold developed and it held its fine aroma and flavor. It grew hard enough to grate, and yet could be sliced very thin.

I called these cheeses a great success.

Hansen's Hard Cheese *

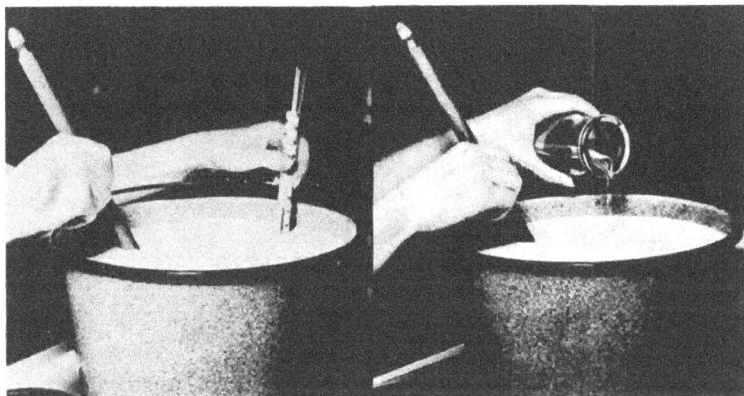

*Left.* Warm the Milk to 86° F.

Use an enameled or tinned pail and heat 8 quarts (2 gallons) sweet whole milk to 86° F. You may use either cow's or goat's milk, with equally good results. If yellow cheese is desired, dissolve an eighth of Cheese Color Tablet in a tablespoon of water and stir into milk.

*Right.* Add Cheese Rennet

Then add ¼ of a Cheese Rennet Tablet (or 2½ "Junket" Rennet Tablets) dissolved in ½ cup cold water. Mix thoroughly. Set in a pan of warm water (85 to 90° F.)

* Directions given by permission of Chr. Hansen's Laboratory.

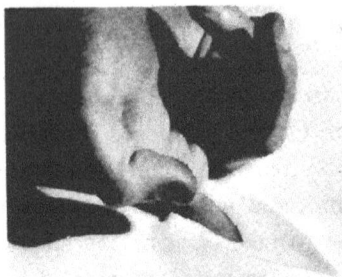

*Left.* Let Set Until "Clean Break"

Let stand until a firm curd forms, about 30 minutes. Test the firmness of curd with your finger. Put your finger into the curd at an angle and lift it. If the curd breaks clean over the finger it is ready to cut. (See diagram *below.*)

*Right.* Warm Slowly to 90–100° F.

Heat the water in the outside pan slowly—allow ½ to ¾ hour to raise temperature to 90–100° F. Stir the curd with your hand very gently at the beginning, so as not to get too many very small pieces of curd. During the entire time of heating, stir frequently enough to keep the temperature even throughout—and to keep the pieces of curd from sticking. Cut with your knife any pieces of curd that are very large; they should all be as uniform as possible.

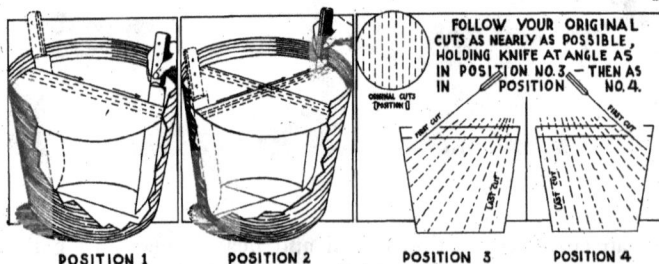

FOLLOW YOUR ORIGINAL CUTS AS NEARLY AS POSSIBLE, HOLDING KNIFE AT ANGLE AS IN POSITION NO. 3.—THEN AS IN POSITION NO. 4.

POSITION 1  POSITION 2  POSITION 3  POSITION 4

Cut 2 Ways Vertically—Then 2 Ways at an Angle.

Use a long butcher knife or pancake turner—long enough so that the blade will go to the bottom of the pail without the handle dipping into the curd. Cut into squares of about ⅜″ (Positions 1 and 2). Use your knife at an angle—(Position 3)—starting about 1–1½″ from side of pail; with angular cuts, slice curd into pieces about ½–1″ thick; begin at top, and make each cut about ½–1″ lower. Turn pail and draw similar angular cuts from other side (Position 4).

*Left.* Pour Curd into Cheese Cloth to Form Round Ball.

When curd is firm enough so it has little tendency to stick together, pour into a cloth about 2 to 3 feet square and form into a
ball. Hang it up until all the free whey has dripped off—2 to 3 hours.
*Right.* Dress the Cheese

Then remove the cloth from the sides of the ball, and place
the ball on a cheese cloth folded over 3 or 4 times. Fold a long cloth,
shaped like a dish towel, into a bandage about 3 inches wide and
wrap tightly around the ball of curd. Pin in place. With your
hands, press cheese down and make the surface of the top smooth
by crumbling with your fingers. There should be no cracks extending into the center of the cheese.

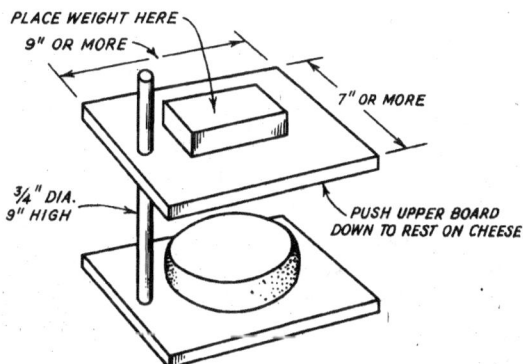

Then Press the Cheese

Lay a piece of wet cloth over the top of the cheese; place a flat plate over the cloth and a weight about equal to a flatiron or a brick. You might find that the weight is likely to fall over to one side, giving the cheese an odd shape; in that case, make a simple cheese press from two boards and a round stick, as illustrated. Your round loaf of cheese should not be more than 6″ across; otherwise it will dry out too much. At night turn cheese and place the weight on top of it again. Let stand until morning.

Store in Cool Place, Salt and Rub

Remove the cloth and bandage and place on a board, if possible in a cool but frost-free place, like the cellar. Turn once or twice a day until a rind is formed. This probably will take three days. Then rub a tablespoonful of salt into the cheese two days in succession. After this rub thoroughly 1 or 2 days with a very small amount of butter; rub and turn the cheese each day until the rind is very firm. After a week or two it will not be necessary to rub so frequently. Two or three times a week will keep it from getting dry and prevents mold from developing.

The cheese can be sold after 3 weeks but will be better after 4–6 weeks' curing at 50–55° F., if stored under proper conditions. A good clean cellar is usually the best place; it should not be so moist that cheese will mold, but on the other hand, not so dry that rind will crack.

## Wisconsin Brick Cheese

Another similar appetizing cured cheese is Wisconsin Brick Cheese, made as follows:

> 3 gals. sweet, whole milk
> ⅓ Rennet Tablet dissolved in ⅓ cup cold water

Heat milk gradually to 86°F. Remove from heat and add ⅓ rennet tablet dissolved in ⅓ cup cold water, stirring thoroughly. Let stand about a half hour, until curd shows a clean break. Cut into cubes ⅜ to ½″ in size.

Heat slowly to 98°F., stirring while heating. Remove from heat and stir with gentle, folding motion for thirty minutes.

Pour off most of whey and work 2 tbsps. salt gently into curd. Place in mold and press for six hours; turn cheese and press for twelve hours.

Rub with salt and store in moderately cool place. In two days rub again with salt.

Turn daily for two to three weeks when it should be ready for use. If mold develops remove with lukewarm salt water.

The mold for this cheese should be of wood, rectangular in shape, like a bottomless box, with holes bored at two-inch intervals to permit whey to escape, and a follower fitted snugly inside and resting upon the curd and weighted with a brick. I bandaged mine in the same manner as the cheese made from the Hansen directions with entirely satisfactory results, although the shape was round, not rectangular.

As I put away the cheesemaking equipment I noted with amusement that except for a dairy thermometer and ply-board for the press, the entire equipment went back onto the pantry shelves. It consisted of:

1 large kettle to serve as a vat into which the milk is placed. This must be of tinware or enamel.

1 larger kettle into which the first kettle is set for heating the milk.

1 dipper with holes for dipping out the curd. Desirable but not essential.

1 tablespoon, 1 teaspoon, 1 large spoon (preferably wooden).

1 long-bladed knife or spatula.

1 good-sized sugar or salt sack or a square yard of unbleached muslin or cotton cloth for draining the curd.

1 cup.

1 colander or strainer.

1 dairy thermometer (This I had to purchase at $1.50).

Smooth boards for making press.

## A Few Precautions

For good results with cheesemaking there are a few precautions to be kept always in mind.

Be sure that the milk from which the cheese is made is clean and wholesome and of good flavor. Don't use old milk.

Have all utensils sterilized.

Follow directions carefully.

Be watchful of temperatures; use your thermometer.

Increase heat gradually.

Stir the curd gently.

Bandage carefully and smoothly, otherwise your cheese may crack.

Use sufficient pressure to expel the whey.

## Goat Meat for the Table.

A MAN who is a butcher as well as a goat-keeper is perhaps better qualified to deal with this subject than many others. Mr. E. T. Harding is both, and this is how he writes to me: —

" I would not be without my goats for all the world. I have five healthy children who have been reared on goats' milk and who are admitted by all who have seen them to be superior, in build, to most others of similar age.

" The present war situation has proved to most people the advantage of keeping goats for their milk supply. We are told by the authorities not to waste a single thing —so why waste unwanted male kids ? These can be used for the table at any age from two weeks. In Ireland they are considered very good at that age. However, they are at their best when four to five months old.

" The quality depends on how carefully they have been reared. If this has been given consideration the meat, when dressed, will look much like lamb, but fatter inside as is the case with goats. The fat is white, and the liver, heart, etc., will be found to be of exceptionally healthy condition and delicious eating.

" The slaughter, from a humane viewpoint, is best performed by a butcher who will, no doubt, be willing to oblige. In addition, he would probably split the carcase, from the tail to the neck and, placing a knife in the tenth rib from the neck, quarter off. The leg is severed from the loin, these forming roasting joints in addition to the shoulder, which is separated from the neck by a vellum.

" The remainder of the carcase makes boiling meat. The carcase should be hung at least a week before it is cut up or it will tend to prove ' rubbery,' as would a fresh killed lamb.

" The kids are best killed before the mating season, as otherwise the meat may be tainted by odour."

I can endorse every word Mr. Harding says, having eaten many a home-reared kid. My local butcher slaughters and dresses the carcase at little more than eighteenpence a time.

## "Gacon."

Then there is "gacon"—or goat meat bacon. Another goat-keeper has written to me as follows : —

"Last autumn I had a very fat 18 months old kid that would not mate, so I got the butcher to slaughter and dress it. My wife cured it as she does bacon, and it is as sweet and tasty as anyone could want. The hams and shoulder pieces we boil, and fry the sides as bacon.

"I am never going to feed an expensive pig again, but always have a winter's supply of gacon. A castrated billy of a large breed at 18 months need not cost much to feed, and makes a good-sized carcase.

"There are probably several ways of curing a carcase, but this is my wife's method:

"Remove the neck and split the carcase down the middle. Cut off the feet and cut out the ham. For curing use a stone slab if possible. For every 20lb. of meat take 1½lb. of salt, 1oz. of saltpetre, and ½oz. Demarara sugar.

"Rub the meat with only the salt and leave with the skin side downward until the following day. Throw away all liquid that has drained out, until the meat stops discharging a day or two. Then apply the saltpetre to the leaner parts, rubbing it in ; also rub in to the whole the mixture of salt and sugar, turning the meat each day and making the mixture last several days.

"When all the mixture is used the meat need not be turned more than once a week. Collect the drainings each day and pour over the meat. Allow the sides and hams to soak 21 days.

"After curing, soak the gacon in chilled water for 15 minutes, then dry thoroughly with a cloth and hang in a draught to dry. When dry, hang in a room not too hot or the gacon will dry up too much, and one not too damp, either.

"Then you can look forward to many enjoyable rashers."

# A Few Recipes.

### GOAT-MILK BREAD.

The following bread is quickly and easily made. It can be kept several days, or it may be eaten hot without fear of indigestion: —

Required: ½lb. plain flour, 1 small teaspoonful bicarbonate of soda, 1 small teaspoonful cream of tartar, 1 breakfast-cupful of goat milk.

Mix the dry ingredients; add milk to make a dough. Put in buttered tin and place in hot oven. Let oven cool slightly and cook for about half-hour, testing with a skewer.

### GOAT-MILK CANDY.

Required: 1 lb. sugar, ¼ pint goat milk, 2 ozs. margarine, ½lb. coconut.

Place in a saucepan, boil fairly slowly and stir occasionally to prevent burning. When the mixture bubbles stir in the coconut. Then turn into flat tins or plates, and cut into squares when quite cold.

### CHOCOLATE CUP.

Take 1 pint goats' milk, 1 tablespoonful cocoa, 1 tablespoonful sugar. Mix cocoa and sugar together. Add 1 pint of cold goats' milk, stirring well. Then pour in 1 pint of boiling water, stirring all the time.

### CAPRA CUSTARD.

Required: Three or four bananas, 1 tablespoonful of sugar, 2 eggs, ¾ pint of goat milk.

Beat the eggs and sugar together in a basin and add boiling milk. Butter a pie-dish and pour this custard in, adding the bananas, which should have been peeled and thinly sliced.

Place the dish in a moderate oven and bake until the custard has set.

## Ginger Goat Cake.

Required : 8ozs. self-raising flour, 4ozs. margarine or goat butter, 4ozs. sugar, 2 teaspoonfuls ground ginger, 1 egg, a little goat milk, and lemon flavouring if desired.

Rub the butter into the flour, mix in sugar, ginger, and flavouring, beat in the egg and milk. Put the mixture into a paper-lined tin and bake in a moderate oven.

## Tea Scones.

Use soured goat milk to make these tea scones:

Take 6ozs. self-raising flour, 3 ozs. butter or margarine, 1 teaspoonful castor sugar, a few sultanas, a pinch of salt.

Mix with sufficient sour goat milk to make into a dough suitable to roll out. Shape and cook 10-15 minutes in a hot oven.

## Night Cap for One.

Take one dessertspoonful of oatmeal, a teaspoonful of syrup, a pinch of salt, cold water, and $\frac{1}{2}$ pint goats' milk. Put oatmeal in a pudding basin and just cover with cold water. Heat the goats' milk to boiling point and pour it into the bowl; add pinch of salt and stir in the syrup. Allow the oatmeal to settle and then drain off the liquid contents into a glass.

# GOAT MEAT

Many inquiries on this subject show the small percentage of our 133,000,000 population of the United States, who are aware of the fact milk goats are like milk cows, i.e. bred and developed for production of milk and not meat. Just as there are beef cattle, so there are some goats which prove satisfactory from a meat standpoint. Those of the common alley class do so and the Angora not only takes on flesh and produces mohair, just as the sheep produces meat and wool. But the only two classes of milk goats which can be counted upon for meat and one is the grade buck kids which should be unsexed at about ten days to two weeks of age whereupon he will grow large and strong under favorable conditions and make delicious meat which can be prepared in the customary way by a good cook. If such stock ranges on pasture instead of dry stall feeding and in fair numbers, it really becomes profitable.

The other class of milk goats which should be

converted into meat is the disappointing milker, for unless she comes from a high producing lineage she probably will not produce daughters of worth as to milk unless said daughters are sired by prepossessing sires. Such does should not be sold to the novice who almost certainly would be disappointed and probably condemn the one who deceived him but he will surely not feel warranted in condemning the entire milk goat industry. One such can destroy the work of a dozen honorable sincere breeders. The disappointing doe should go to the meat block and not be permitted to defame her better sisters.

# ANGORA VENISON.

NGORA venison is the name which should be given to the flesh of the Angora goat. At the present time it is usually sold in the markets as mutton. The term goat meat should be applied to meat of the common goat, and the term mutton belongs to sheep. Because the Angora goat feeds largely upon that material which nourishes the deer, the meat of the Angora is flavored like venison. The fat is well distributed, and the healthfulness of the animal renders this an especially desirable meat. The Turk has long recognized Angora venison as an important element in his diet. Angora kid is above comparison, and it occupies the principle place on the menu at private as well as state affairs in the Orient. As one passes through the market places in Asia Minor he sees the carcasses of the Angora hanging in every shop. There is no mistaking the animal, as the skin still remains on the goat. One takes his choice, and as a rule more Angora venison than mutton is sold. Some of the Turks keep their wethers until they become coarse-haired and too old to pay to keep longer, eight or ten years old. This class of meat ranks with old mutton, and sells at a discount. Young wethers and does are in good demand. There has existed in America some prejudice against the flesh of the goat. To-day thousands of goats are being consumed annually, but most of them are sold as mutton. Packers and butchers still insist that Angora venison must be sold as mutton. They pay about one-half a cent to a cent a pound less for the goat than for sheep.

The goat never fattens as well along the back as the sheep, and hence the carcass does not look so well. The fat is more evenly distributed throughout the animal in the goat. An expert once said that to know whether a goat was fat one should feel the brisket, and if there was a considerable layer of adipose tissue between the skin and the breast bone, the animal was fat.

Some of the American breeders do not send their wethers to market until they get too old to produce valuable fleeces. The animals are then slaughtered when they have grown a half year's fleece, and the skins are reserved by the breeder. These skins are valuable, and help to bring up the average price of the goat.

At present some of the packers recognize no difference between shorn and unshorn goats. The price is the same, so it pays to shear the goats before bringing them to market. There is absolutely no strong flavor in prime Angora venison, and this is where the meat differs from that of the common goat.

The goat is a slow grower, and not until the second year do the bones ossify. Therefore, a two-year-old can be sold for lamb, as he has a "soft joint." Grown Angora wethers do not average much more than one hundred pounds as a rule, although there are occasional bands sold which average one hundred and fifteen pounds.

It is safe to say that Angora venison will never supplant mutton, but it will have its place among the edible meats.

ANGORA BUCK—Early Importation.

# ANGORA GOAT SKINS.

N Angora goat skin differs considerably from the skin of the common goat. In the first place the Angora skin is covered with more or less mohair; and in the second place, the texture of the skin itself is different. The skin of the common goat is firm, and the different layers are so closely united that they cannot be separated. The layers of the Angora skin are not so closely united, and the skin is slightly fluffy. The outer layer of this skin peels off when it is used. The Angora skin is valuable both with the fleece on and without it. Its principle value, however, is with the fleece on. After the skins have been properly tanned, they are used for rugs, robes, trimmings, and imitating various furs. When ladies' and children's Angora furs are in style, these skins become very valuable for this purpose. One skin has cut $17.00 worth of trimming at wholesale. Of course, the value of the skins depends upon the quality and character of the mohair with which the skins are covered, and their size. Large, well covered skins are always scarce and command good prices. They are worth from $1.00 to $2.00 each.

Most of the Asia Minor skins are sent to Austria, and the prices paid for the raw skins are about the same as in America. The skins which have had the mohair removed are valuable for the manufacture of gloves and morocco leather. They do not make as fine leather as the common goat skins, but they are as extensively used. All skins should be carefully handled.

The skin should be carefully removed from the carcass. Goats do not skin as easily as sheep, and the careless operator is liable to cut the inner layers of the skin if he is not careful. These cuts are called "flesh-cuts," and skins badly "flesh-cut" are comparatively valueless, because "flesh-cuts" can not be removed by the tanner. A sharp knife should be used, and the operator should avoid cutting the skin.

The skin should be well salted, care being taken to see that the salt penetrates every portion of the raw surface. The skins can be cured in the shade without the use of salt, but sun-dried skins are worthless. If the edges of the skin are allowed to roll, so that raw surfaces come together, the part so affected will heat and the hair pull out. It is not necessary to stretch the skins while curing them.

Goats should be killed when their fleece is suitable for robe and rug purposes. Those carrying a six month's fleece, if it is six inches long, have about the right kind of skins. There are some Angora skins imported from Turkey and South Africa.

Prize winners at the Columbian Exposition, Chicago, 1893.

# BY-PRODUCTS
## OF ANGORA GOATS.

HE Angora goat should not be classed with milch animals. As a rule the does give a sufficient amount of milk to nourish the kid or kids. The more common blood there is in the goat the better milch animal she is. However, some Angoras have been milked, and the milk is as rich as that of the common goat. A quart of milk a day may be considered a fair average for a fresh milch Angora doe. It has been suggested that because the milk of the goat contains a heavy percentage of fat, it is a proper substitute for mothers' milk for babies. This is probably a mistake, as that part of the milk which is the hardest for the baby to digest is the protein, and it will be observed that in the following table of analysis submitted, the percentage of protein in goat's milk and in cow's milk is about the same, and that it is considerably larger than in mother's milk. A very desirable feature in goat's milk is that the fat is distributed throughout the milk, and that it does not readily separate from the milk. This would assist in the assimilation of the fat by an infant. Some experiments made with coffee demonstrate that it requires half the quantity of goat's milk to produce the same effect upon this beverage which cow's milk produces. This may be partially explained by the quantity of fat in goat's milk, and partially by the fact that the fat does not readily separate from the milk. The bottom of the can is as good as the top.

## ANALYSIS OF MILK.

|  | MOTHER'S AVERAGE | COW'S AVERAGE | GOAT'S AVERAGE |
|---|---|---|---|
| Fat | 4.00 | 3.50 | 7.30 |
| Sugar | 7.00 | 4.30 | 4.10 |
| Proteid | 1.50 | 4.00 | 4.18 |
| Salts | .20 | .70 | 1.21 |
| Water | 87.30 | 87.50 | 83.21 |
|  | 100.00 | 100.00 | 100.00 |

Persons in poor health have been greatly benefitted by the use of goat's milk. This is probably due to the fact that the fat in the milk is so distributed that a large percentage of it is taken up by the digestive apparatus. Angora goats are docile, and it is possible that some of them could be developed into good milch animals.

## FERTILIZER.

It is a known fact that packers of the present day utilize all of the carcass of most food animals, but it is not the fertilizer which the packer makes from the blood and offal of the goat which we shall consider here.

Sheep's manure has been used for years on orchards and vegetable gardens, and in the last few years goats' manure has been in demand, selling at from $6 to $7.50 a ton, depending upon the purity of the fertilizer. It must be remembered that only a small portion of this manure is dropped at the night bed-ground, the balance is evenly distributed over the land upon which the goats are feeding. The goats not only rid the farm of objectionable weeds and brush, but they help to furnish a rich soil in which grass will grow. This fact has been so thoroughly demonstrated that western farmers, who have large tracts of wheat or barley stubble to rent during the summer, are always anxious to get goats upon this land.

## OTHER PRODUCTS.

The horns of the goats are used to make handles for pocket knives, etc. The hoofs are used in the manufacture of glue.

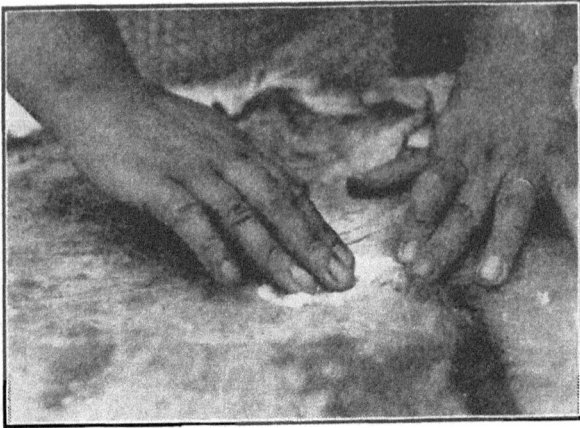

The skins of unwanted kids can be made up into many beautiful articles.
The curing process is simple, the skin being treated with a mixture of
alum, salt and saltpetre.

After treatment the skin is wrapped tightly in newspaper and put away
for ten days or so. For full curing instructions see page 45.

### Curing Kid Skins.

I HAVE known goat-keepers who, in disposing of unwanted kids, have disposed also of the skins. This may be an excusable extravagance in peace—at the present time, with the national call to economise, it must not be countenanced.

Many beautiful articles can be made from kid skins, and it is perhaps needless to remind lady goat-keepers of the fur coats, jackets, capes, neckpieces, gloves, baby carriage covers, rugs, and so on, which can be made up.

There are several methods of curing skins, but I learned the following simple formula some years ago, and it has always proved successful: —

The skin to be treated must be as free from fat as possible. Use a blunt table knife to scrape off any that is adhering, the skin being placed fur down on several thicknesses of paper.

The curing mixture is equal proportions by measure of salt, saltpetre, and powdered alum, say about a tablespoonful of each for one or two small skins.

Rub this mixture well into the fleshy side of the skin, using the balls of the three middle fingers. Take care in going round the edges. Should the mixture get into the fur it will cause sweating.

Having been all over the skin, place three or four folds of paper on top, roll up tightly, and put away the package for 10-14 days. Then take out the skin and, with the aid of a blunt knife, peel off the papery tissue. This should come away quite easily, exposing a soft, pliable chamois-like skin.

Another method is to tack the skin to a board, fur down, and paint it once every day for a week with alum solution, finally rub well with emery paper, and then work the skin vigorously between the hands. The skin is then ready for use.

In making up, skins should never be cut with scissors,

or the fur comes out. With the skin laid fur down on a flat surface, and the pattern marked out with a pencil, cut fairly lightly with a sharp penknife. If you cut heavily you will go through the fur. I understand that No. 30 sewing cotton is the best for kid skins. Edges should be over-sewn carefully, too deep stitches or pulling tightly on the cotton should be avoided. Any flaws or thin places in the skin can be cut out and a good piece of fur sewn in.

If readers wish, I can advise them of a firm which specializes in the curing of goat-skins at reasonable prices.

## STARTING IN BUSINESS

No sane person contemplating the establishment of a new business in any line would be so rash as to invest every dollar available without first becoming to some extent familiar with the various intricacies of that particular line. The sign posts to success in this particular activity point to two ways for gaining knowledge. One being practical experience under the guidance of a reliable and successful breeder or dairyman giving preference to the line preferred or a careful study of the writings of several accepted authorities. By no means permit the various ideas, plans and formulas advocated by some conflicting with others to cause you to conclude because all do not agree, all must be wrong, for it is worth bearing in mind some of the most learned men of the top professions disagree. A line of reasoning based upon the high points advocated should serve as a safe conclusion as not being far from right, until your own experience demonstrates which is best. The first step is to decide which branch of the Industry shall have your foremost effort. Whether as a breeder of high bred stock for sale, for breeding purposes, or for exhibition purposes or for dairy purposes, or just

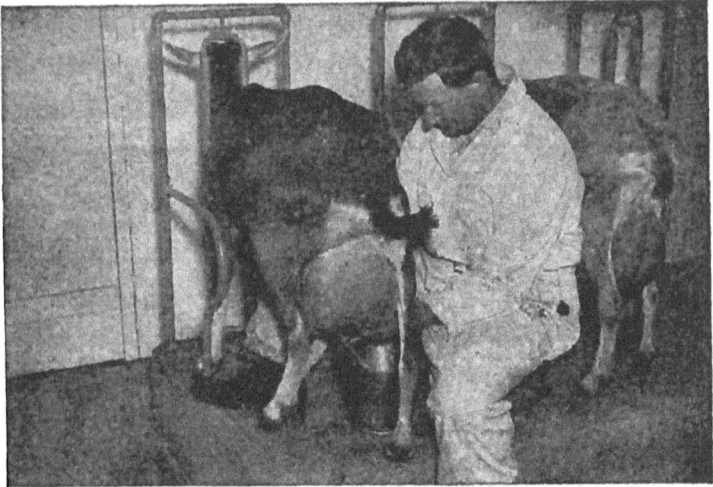

COMMONLY PRACTICED IN SMALL PRIVATE STABLES

for utility purposes. In any event your decision as to breed is paramount and that same question arises in the mind of the beginner whether it applies to horses, cattle, sheep, swine, poultry or milk goats and it is highly important the one which appeals most is the one to select, for any business man succeeds best in the line which he likes best.

Likewise, it is very important to determine the

location of your activities in keeping with your aims, deciding upon space, improvements and numbers to fit. One can start on a small scale by

LOVELY INNOCENCE

selecting a first class pure buck of the breed of his choice and good registered grade does of the same breed raising the doe kids which each generation step up in percentage of blood lines, always destroying the grade buck kids or emasculating them when ten days of age and growing them into meat

PRACTICAL INTERIOR PLAN

REAL CONTENTMENT

# IMPROVED MILK GOATS

## TABULATED PEDIGREE OF

Name        Sex        Breed        Size

Born    Color    Horned or not    Breeder    Owner

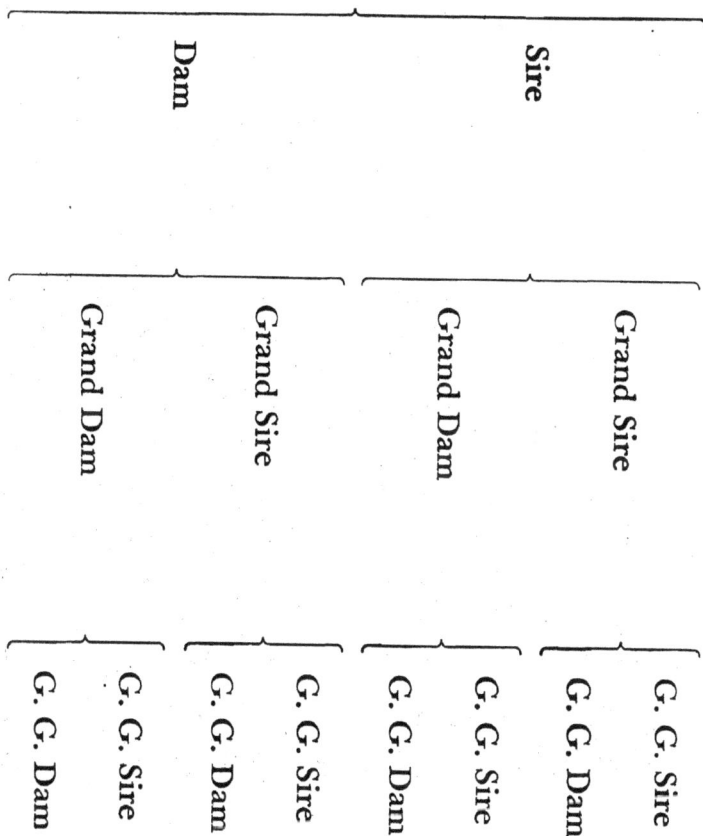

provided you do not sell milk and have a surplus which otherwise would be wasted.

The first cross of a native scrub doe by a pure registered buck produces ½ blood doe kids, the second ¾ the next ⅞, then 15/16 and the fifth

PLAYFUL EXERCISE

cross 31/32 which from there on are known as American Toggenburg, Saanen, etc. When you have attained the 10th cross which is 1023/1024 they are accepted as pure and the 10th cross is far more pure than some other lines of live stock require for registration as pure. It is surprising how in some extreme instances departing from color, may give a shock to the breeder who keeps but one breed with no possible opportunity for crossing with a buck of another breed. This was demonstrated in the earlier importations and we felt possibly the foreign breeders had been neglectful of their fencing or housing but with all our surmises it must be borne in mind these are what one commonly terms "throw backs" and even they may and usually do produce in accordance with the regular type of their immediate ancestors.

Success may come more quickly to the one who has capital coupled with business experience in other lines whereby judicious management counts, but nothing counts so much in production as proper care and that is a subject for further consideration.

# SELLING MILK

Where can I sell my surplus milk, is a question which I presume stands second only to the inquiry as to which breed I prefer. It is useless to formulate a long procedure, but a fairly flexible rule should meet the requirements in most localities. When and if no previous effort has been made to introduce goat milk in a community, large or small, it is practical to inform the public of it being produced in a strictly sanitary place and manner, also that it can be obtained at a fair price, and just here. What is a fair price? The answer is cost of production and delivery plus a reasonable profit and by no means a price which will prove prohibitive, soon as the immediate need of it is past. The most practical and economical way in addition to modest advertising in your local newspaper is to supply your public library, lodge and club, reading rooms with the best periodical on the subject, and leaflets which are obtainable at cost from the publisher of said periodical. This will

set the wedge and the testimonials of those who receive beneficial results from its use will drive it
home to those who at first are indifferent or prejudiced, for a decidedly large percentage of citizens
of this country are not aware of the life-giving vitality in each and every drop of nature's finest
food. Oh, yes, you will be confronted by long dis-
sertations in various forms. Some will be by those
who actually fear competition and who forget the
old slogan "Competition is the life of trade," while
others honestly believe the jokester and cartoonist
have the correct slant, but just let him or one of
his family reach a point where numerous efforts of
well meaning attendants have failed and some sane
person, who has had the practical experience of
using the most easily digested and most nourishing
food, pound for pound, known, comes along and
insists upon it being given a fair trial, and the result is the support of added converts.

Strange when practically all the older countries
have long since adopted the goat as the most
cleanly, healthy and affectionate one of all the
barnyard, that our country, which has set the
world agog with our accomplishments in almost
every other line, should be so slothful in awaken-
ing to a fact so important to our welfare.

There are various plans for marketing the milk and here given is the order of their most common adoption. One is to deliver direct to your customer in quart or pint bottles, which includes time and distance which must be taken into consideration. Second, when your cow dairyman realizes he can just as well hold his customers who have temporarily ceased to patronize him in order to use goat milk, and, if he is a good business man, he will soon see the advisability of arranging with you for a stipulated daily amount at a figure he can afford to pay. After adding cost of delivery and a slight profit, he will either take the initiative and approach you or at least give consideration to your proposition, if you will show him first that you are not fighting him nor the cow in general, nor can you afford to produce and deliver the milk as cheaply as to sell to him in bulk. Third, then in some of the larger cities, there are distributing centers owned or controlled by people of sufficient capital to ride the waves of introducing what at first they may term an experiment, and who along with their regular means of advertising can at a nominal additional cost offer your milk to their own patrons, both retail and to the dairyman, who from time to time, due to shortage may go to such centers and purchase an amount sufficient to tide him over until his own production is restored. Fourth: A most novel and apparently successful way is one adopted in one of the first cities of the land and which impressed me as being practical after inspecting the farm, to find a strictly modern system and equipment approved by the Board of Health, which ordinarily does not overlook minute detail; the milk was delivered in pint bottles to fourteen groceries and delicatessen shops where the milk goat dairyman had installed electric refrigerators, occupying a prominent position on the floor. He left a supply governed by the past requirements, taking up the unsold bottles and the entire transaction was based upon a saving of time to not only him, but to the grocer as well, for the customer necessarily must deposit the coin before the door would open, whereupon he could help himself to the amount paid for (true he was upon his honor as to the number paid for and taken) and the grocer busy with others had lost no time and received his compensation weekly by way of a percentage of net sales as shown by the cash drawer. Then too, his customers could have a novelty, if so regarded, or a trial of what was new to them and the dairymen's sales climbed and expense declined and profits increased.

Still another system rather in its infancy, is the use of thermos bottled milk, which formerly went through the regular mails with rigid requirements as to packing in order to avert spilling and damage to other items exposed to the consequent dampness, but now it is being more and more the custom to send via air at slightly increased cost of transportation but far better results gained by prompt service. Moreover it is claimed in some instances just in time to save much suffering or even a life and what would we not do to accomplish that?

# SALE OF STOCK

As every manufacturer, wholesale or retail dealer knows, you must have what is wanted and at the right price in order to sell. Here is where a breeder must determine along which line he proposes to devote his attention. If stock is to be raised for exhibition purposes, then it should be from winners in the show ring and it is well to have kept a list of dates and places where prizes have been won by the ancestors of your sale stock, which can easily be compiled by noting the winners in reports of fairs and livestock shows regularly. File those reports and for your convenience it is well to index them for reference because you may have occasion to refer to them in connection with some new stock you may own later and which are not related to the winners you had previously listed.

## Special Stock

If stock is to be raised and sold for dairy purposes, the breeder is not necessarily devoted to the reports of the show ring so much as to the reports of milk contests and I have frequently made the statement. If our milk goats are not producers of milk in paying quantities, why call them milk goats?

Then there is the breeder fortunately situated with broad acres for pasture and in some climates favorable to many months of such inexpensive maintenance, only calling for a comparatively small amount of grain, who can produce good utility stock within the reach of the average family in limited circumstances. And what a blessing it is to be able to render service there, for without you, it might be they would never have the advantage you can give them. Summing up sale of stock of all branches and the products from some, there is no better guide than to use good sane, honorable practices and you will find others will recommend you, and what is of far greater importance, you will have the right to feel you have served faithfully.

# Goats as a Business

MORE or less frequently one meets the question, "Is there money in it?" It is difficult to give an answer that the questioner will understand, but it would seem that if your chief purpose in having goats is to make money you would do better with less effort in some other field. There are people who find goats profitable, modestly so, either in combination with some other activity such as chicken raising, or as a full-time venture, but they are people who, first of all, have goats because they love them. In addition to the return in dollars and cents they value the less tangible compensation such as comes, for example, in an afternoon in the woods with the herd, and the feeling of quiet peace it brings.

If you think of going into goat raising as a business be very sure that you have the temperament for the work. It requires, above all else, limitless, untiring patience. Plenty of people have started goat raising only to give up in a short time. They will tell you that the work is too hard; the goats require too much care; it's too confining. All this is true, but if you like it you can omit the "too." Like a physician you are on call twenty-four hours of the day seven days of the week. Goats are naturally exceptionally healthy animals, but it is your care and watchfulness that must keep them healthy. The routine work goes on winter and summer, in good weather and bad, and it must be carried through on schedule or you will run into trouble. During the kidding season it is often necessary to spend the entire

How many milkers are needed for a family to make a living? One successful family dairy has 25—plus a retail route.

night at the barn, and it may be a very cold night. Next day you can't rest and make up the lost sleep, but must carry on as usual.

If members of the family can share the work it makes the going easier, but if outsiders must be employed you still, as owner, must bear the care and worry and keep the ever-watchful eye, for to secure a worker who will have the same interest in your herd as you have is practically impossible—too much to ask of fate. Even if you are worth a million you can't necessarily get the right sort of person to take over. In fact at one time a millionaire goat owner told me he was disposing of his herd of 300 or more because he could find no one to take proper care of them. When my herd was small I boarded them

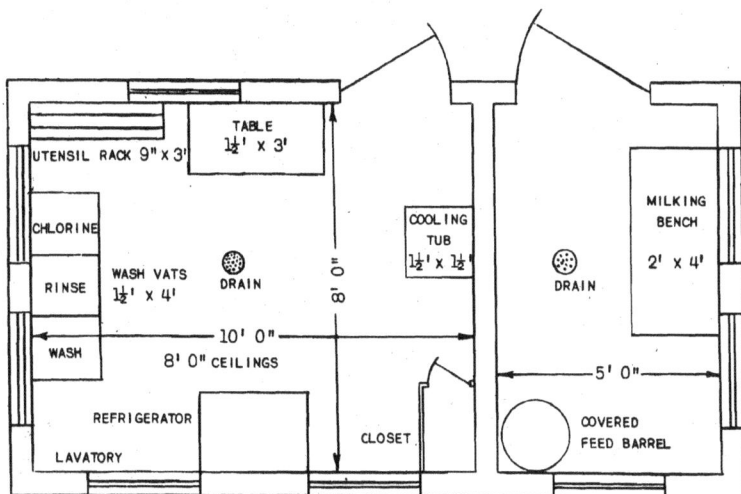

Plans for a "Grade A" Goat Dairy Milk House.

Specifications

All floors of smooth concrete slope to drains; rounded corners 6 inches about floor level. Walls and ceilings may be painted plywood. All openings screened with 16-mesh wire. Doors self-closing to open outward. Running water must be piped to the wash vats. Lavatory may consist of wash basin and paper towels. Drains with bell traps; piped 50 feet from building. Steel table. Gasoline or gas heater for heating water, or laundry stove. Tub with ice water for cooling to 50° F. If refrigerator is in satisfactory condition, it may be kept in owner's house.

Utensils

Covered milk buckets—three-quarter top. Metal strainer with cotton pad—no cloth. One valve bottle filler—no pouring. Mechanical hand capper.

Caps purchased in tubes and kept in a clean, dry place. Name of producer and contents printed on cap.

for a few months with a man who considered his herdsman excellent. One day I stopped in at feeding time. My two six-months-old kids were together in a box-stall, which was nice for companionship, but they had one feed box between them and the more aggressive kid got the food which was dumped into the box on top of the stale, left-over bits and a few sprinklings of manure from the kids' hoofs. These are the conditions that the watchful goat owner would guard against.

If you are primarily interested in dairying, be sure that you are accessible to customers, both for their sake and for your own, for your location will do much to advertise your dairy, and though goat milk may be as desirable as that proverbially superior mousetrap, customers won't blaze a trail to your dairy to obtain it. Acquaint yourself fully with the state and local requirements for selling milk. Recognize that selling goat milk is a job that requires constant effort. It isn't like the sale of cow milk. Every family uses cow milk and when a new family comes into a neighborhood it is only a question of which milkman gets there first—the customer is waiting. With goat milk you will find that someone's baby will need the milk and then as the child progresses the mother will one day tell you "I'm going to put him on cow milk now." Your customer is gone and you must find another to take his place. Some days you won't have enough milk to supply the calls and you will begin to wonder if you should expand, then a slump will come and you will think of the mounting feed bills and perhaps sell off some of your stock only to regret it later. If you raise your own hay and grain this is a lesser worry, but even so, you don't like to lose customers.

Another important matter to keep in mind if you are dairying is a steady flow of milk. Many customers are lost because the dairy has to post a sign or notify its customers some time in the late fall or winter months—"No milk available until January." For customers allergic to cow milk or babies on a goat milk diet this is serious, and it behooves you to arrange your breeding schedule so as to avoid so far as possible this shortage. Make a list or chart of your does and plan their breeding as follows:

As early as possible, in August or September, breed the yearlings and does whose lactation is slowing down, provided

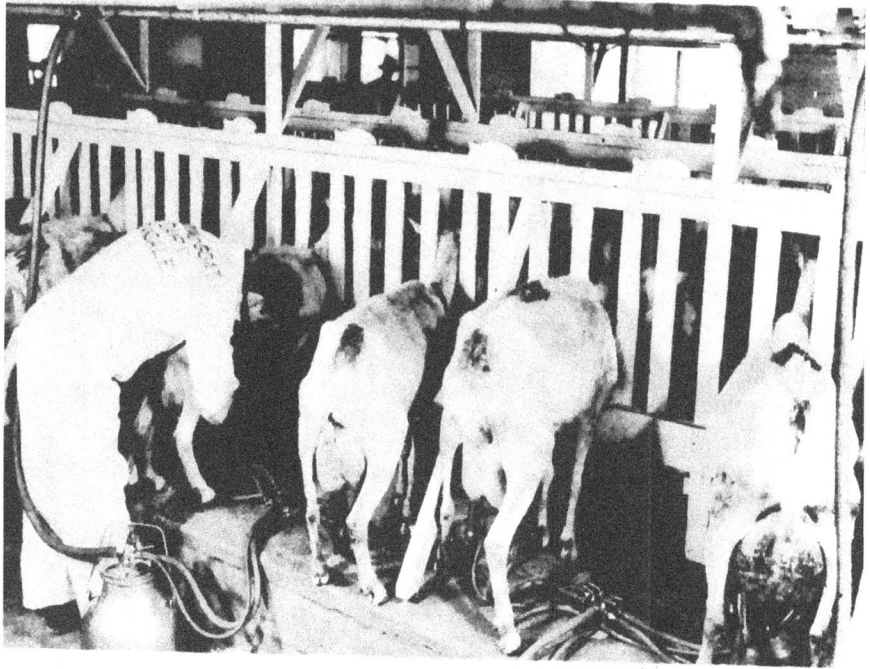

The main advantage of a milking machine is in speeding up the milking—and it's easier to buy a milking machine than find a good milker.

those does have not freshened too recently. To maintain their good health and vigor they should not be expected to kid more than once a year.

In October and November breed the next group, saving for December, January and February breeding your best milkers —those with longest lactation period.

In this way you will have does freshening in January and February, March and April, May, June and July. Presumably some of your does, even though pregnant, will continue to milk for at least three months after they are bred and the late fresheners will carry you through the winter months until the first group freshens. Although the breeding season extends into March it is a little risky to hold any does so late as you have less chance to repeat the service in case it is not effective.

Be very sure that you supervise this breeding schedule yourself. During October one season it was necessary for me to be away for a week and I gave instructions that one doe who had freshened in July should not be bred. She was a Nubian with distinct individual markings and there should have been no misunderstanding, but on my return I found that she was the only doe that had been bred. At kidding time she died. Recently a customer who had purchased a young doe wrote that she was disappointed in the production of Daisy. It astonished me that Daisy had been bred at just six months old, and I learned that the owner was away when Daisy came in heat and the hired man had bred her.

The breeding end of the goat business seems to be the more popular. For one thing a goat breeder can locate on cheaper land, provided he is accessible to a railroad station for shipping stock. He has less routine milking and kid feeding to do as he can leave the kids to suckle from their mothers. He doesn't worry about the loss of milk customers or hustle about to find new ones, and any excess milk he can convert into butter or cheese for his own use. To be sure he must find customers for his stock, but that isn't difficult if he has good animals and advertises in the right publications.

During the war, which prevented the importation of foreign cheeses, a very active interest developed in the making of cheese from goat milk. In some cases a herd is maintained to provide milk and, in addition, milk from dairies is purchased. A business of this sort calls for quite some investment in cheesemaking equipment and the employment of skilled workers to make the cheese. If you are located near such a plant you might find it profitable to arrange with them to take over your surplus milk, provided you can meet their requirements.

On the Pacific Coast, at Soledad, California, the Meyenberg Milk Products Company, originators of the evaporation process for cow milk, have a plant for the evaporation and canning of goat milk. This sort of venture also requires heavy investment and full knowledge of how it is done.

Tie stalls in a commercial dairy.

There are also some people who have gone into the cosmetic field, making creams and lotions from goat milk and whey. If you have a knowledge of chemistry this field might bear looking into, as very little has been done in it thus far.

Whatever branch of the goat business you undertake be sure that your barn, however simple, is clean and attractive, and your animals also. Have your equipment as complete as you can afford so that you won't suffer the nervous strain and the extra work that comes with makeshifts.

www.ingramcontent.com/pod-product-compliance
Lightning Source LLC
Chambersburg PA
CBHW061235220326
41599CB00028B/5430